Postharvest Management and Value Addition

THE EDITORS

Dr. Ashwani Kumar Goel is currently working as Head, Department of Agricultural Engineering, Hamelmalo Agricultural College, Keren, Eritrea (Africa). He obtained his B.Tech (1981) and M.Tech (1985) from Punjab Agricultural University, Ludhiana and Ph.D (2002) from G.B. Pant University of Agriculture and Technology, Pantnagar. In the field of Agricultural Engineering, he has 20 years of experience in teaching, research and extension in the CSKHP Agricultural University, Palampur. As a scientist, he has visited and presented his research work at Israel, Belgium and Thailand. He is a Fellow of the Institution of Engineers and life member of Indian Society of Agricultural Engineers.

Er Rajinder Kumar is presently posted as Sr. Agril. Engineer at CSKHP Agricultural University, Palampur. He did his B.Tech. (1969) and M.Tech. (1971) from Punjab Agricultural University, Ludhiana. He has served the cause of the profession for more than 34 years in teaching, research and extension in different capacities. He is a member of several professional societies and remained on the editorial board of many journals of repute. He has more than 100 publications to his credit in various national/ international journals. He has attended many conferences/workshops. He has handled many research projects as principal investigator in collaboration with different agencies. His research work over the years has benefited the hill farmers in particular.

Er Satwinder Singh Mann is currently working as Lecturer, Dept, of Agricultural Engineering, Hamelmalo Agricultural College, Keren, Eritrea (Africa) He possesses more than five years experience in project handling and teaching in the field of Farm Power and Machinery, Agricultural Process Engineering and Renewable Energy Sources. He has obtained B.Tech (1992) and M.Tech (1999) from Punjab Agricultural University, Ludhiana.

Postharvest Management and Value Addition

— Editors —

Ashwani K. Goel
Head, Department of Agricultural Engineering,
Hamelmalo Agriculture College,
Eritrea (Africa)

Rajinder Kumar
Sr. Agricultural Engineer,
CSKHP Agricultural University
Palampur (H.P.), India

Satwinder S. Mann
Lecturer, Department of Agricultural Engineering,
Hamelmalo Agriculture College,
Eritrea (Africa)

2016

Daya Publishing House®

A Division of

Astral International Pvt. Ltd.

New Delhi - 110 002

First Published, 2007
Reprinted, 2016

ISBN: 9789351306481 (International Edition)

Publisher's note:

Every possible effort has been made to ensure that the information contained in this book is accurate at the time of going to press, and the publisher and author cannot accept responsibility for any errors or omissions, however caused. No responsibility for loss or damage occasioned to any person acting, or refraining from action, as a result of the material in this publication can be accepted by the editor, the publisher or the author. The Publisher is not associated with any product or vendor mentioned in the book. The contents of this work are intended to further general scientific research, understanding and discussion only. Readers should consult with a specialist where appropriate.

Every effort has been made to trace the owners of copyright material used in this book, if any. The author and the publisher will be grateful for any omission brought to their notice for acknowledgement in the future editions of the book.

Published by : **Daya Publishing House®**
A Division of
Astral International Pvt. Ltd.
– ISO 9001:2008 Certified Company –
4760-61/23, Ansari Road, Darya Ganj
New Delhi-110 002
Ph. 011-43549197, 23278134
E-mail: info@astralint.com
Website: www.astralint.com

Preface

It gives me immense pleasure to add preface to this book on Postharvest Management and Value Addition. Alongwith the 'information explosion' in the scientific world during the recent past, there has been rapid and enormous progress in science and technology concerned with postharvest management of agricultural/ horticultural produce. Latest advances are fast transforming postharvest management from a means of subsistence to an organized industry. This has stimulated a great deal of interest among widely diversified group of scientists. In this era of high yielding crop varieties and multiple cropping techniques, global competition in the quality of food products has put every country on the alert to improve upon postharvest techniques and value addition sector. In India, the processing of only 2 per cent of the total produce of fruits and vegetables is at present undertaken as compared to 70-80 per cent in the developed countries. However the Government of India has fixed a target of 10 per cent of such produce to be processed at the earliest. Also, the actual producer is not getting his due share from the prevailing market prices.

The ever-increasing scientific research on the subject has resulted in a large body of knowledge which can be found scattered in various scientific journals and technical reports and is not easily accessible. The available material needs to be assembled and briefed in order that the students/technologists/agriculturists/ horticulturists/engineers and others actively engaged in quantitative studies can have it as a ready reference. This book provides a comprehensive but concise treatise on the subject matter. The text consists of 15 chapters which cover different aspects of postharvest management. The subject requires inter-disciplinary approach and cannot be dealt properly without any joint venture. The contributors to this volume are from the faculty of different disciplines specialized in their respective fields and the manuscript covers a wide range of postharvest operations right from handling to management of harvested grains/fruits and their value addition.

I sincerely appreciate the contribution of the authors, editors as well as the publisher of the referenced literature for making such useful information available to the scientific community. The great effort on their part will find favour with the teachers and the students alike and will impart the extension workers with necessary impetus to undertake the desired postharvest management practices. The book in its own way endeavors to help in the development of postharvest technology in India and create employment opportunities.

Dr. D.S. Rathore
Vice Chancellor
CSK HPKV,
Palampur (H.P.), India

Prologue

Postharvest management is an emerging area in the present scenario encompassing a widely diversified group of scientists. India has now attained the stage where apart from enhanced production of crops, emphasis is being laid on processing and value addition of cereals, vegetables and food crops. However, the utilization of fruits and vegetables for processing in the organized and un-organized sectors is around 2 per cent of the total production. Given the huge potential for preparation of processed food products and limitation of the domestic market in consuming these, it is imperative that concerted efforts are made to export processed food for which the private, public and co-operative sectors, all have to play their rightful role. Value addition is not only a science, it is an art too.

A single comprehensive book clearly written and well laid out is rare. This manuscript deals with multifarious aspects of postharvest technology and contributions have been made by the distinguished scientists par excellence in the respective field of specialization. Voluminous material on the subject has been presented in a concise way so that the utility owners can have it available as ready reference. The practices described are mainly those prevailing or coming into use in India and the approach to the subject has been practical through out. Useful research data and illustrations make the comprehension of the subject easy and the language is clearly understandable by all concerned.

The editors have derived unequivocal gratification from the advocation by the Hon'ble President of India, A.P.J. Abdul Kalam to join the band of developed world which has a major share in export due to the adoption of advanced post harvest techniques. We are gratefully indebted to Dr. Nawab Ali, DDG (Engg.) for contributing a chapter on soybean food potential and Dr. Pitam Chandra, ADG (Process Engg.) ICAR, New Delhi for constantly encouraging the compilation of this document. We owe to all who rendered help in different ways and are of the opinion that the submissions made by the different authors are genuine.

Errors and shortcomings if any may be debited to the editors and credit to all the authors. Suggestions from the readers for improvement in the book will be most graciously acknowledged.

Ashwani K. Goel

Rajinder Kumar

Satwinder S. Mann

Contents

Chapter 1

Soybean Food Potential and Technology for its Utilisation in India

Nawab Ali

Director, CIAE, Nabi Bagh, Berasia Road, Bhopal – 462 038

Soybean is a food legume and an excellent source of nutrition (Table 1.1) and health promoting phyto-chemicals at an affordable price. It has been reported that the regular use of soybean in daily diet enhances and protects human health and results in longevity– the goal every human wants to achieve. Hence, the direct food-use of soybean is in the interest of mankind and it, therefore, should be promoted globally and particularly in India where a majority of its population is vegetarian and suffers from protein-energy malnutrition.

Production and Productivity of Soybean

Soybean was introduced in India from China through Himalayan mountains several centuries ago. Now, it is cultivated in about 65 million ha in India and has given birth to a chain of soybean-based industries. There has been a significant improvement in socio-economic status of soybean farmers, particularly in Madhya Pradesh. Soybean production in India has decreased from a peak production value of 6.95 million tonnes in 1998-99 to the level of about 5 million tonnes in 2002-03 and 6.93 million tonnes in 2003-04 (Table 1.2). The current minimum support price for soybean is Rs 8850/t. It is expected that in 2004-05, soybean production in India would be about 7.0 million tonnes.

Table 1.1: Approximate Food Value

Constituent	*Food Value, 100 g of Edible Soybean*
Protein	43
Carbohydrates	21
Fat	20
Moisture	8
Minerals	5
Fibre	4
Energy, kcal	430

Source: ASA, 1999 and Gopalan, 1974.

Table 1.2: Production, Area and Yield of Soybean in India During 1970-1971 to 2003-2004

Year	*Production, million tonnes*	*Area, million hectare*	*Yield, kg/ha*
1970-71	0.01	0.03	334
1975-76	0.09	0.09	1000
1980-81	0.44	0.61	721
1985-86	1.02	1.34	761
1990-91	2.60	2.56	1015
1995-96	5.10	5.04	1012
1996-97	5.38	5.45	987
1997-98	6.46	5.99	1079
1998-99	6.94	6.31	1100
1999-00	6.79	5.98	1135
2000-01	5.35	6.15	870
2001-02	5.20	6.00	865
2002-03	5.00	5.60	890
2003-04	6.93	6.45	1074

Source: Agriculture Statistics at a Glance, 2001 and Singhal, V, 2003.

The annual soybean production in India has been about 6 million tonnes with an average yield/productivity of 1000 kg/ha.

Madhya Pradesh is the major soybean growing state of India contributing about 60.1 per cent of the total soybean production (Table 1.3) followed by Maharashtra (28.2 per cent), Rajasthan (8.5 per cent), Andhra Pradesh (1.1 per cent) and others (2.1 per cent). Soybean contributes about 12 per cent to the edible oil pool of the country (Table 1.4). The first two major edible oils of India are rape/mustard (36 per cent) and groundnut (12 per cent).

Table 1.3: State-wise Production and Productivity of Soybean in India During 2003-04

State	*Area, million hectare*	*Production, million tonne*	*Yield, kg/ha*	*Share in Total Soybean Production, %*
Madhya Pradesh	4.090	4.169	1019	60.14
Maharashtra	1.560	1.954	1253	28.19
Rajasthan	0.563	0.595	1057	8.58
Andhra Pradesh	0.082	0.077	940	1.12
Karnataka	0.060	0.054	900	0.78
Chattisgarh	0.050	0.042	846	0.60
Others	0.050	0.041	830	0.59
Total	**6.455**	**6.932**	**1074**	**100.00**

Source: Soy-update, 2004.

Table 1.4: Production of Edible Oils in India During 1999–2000

Edible Oil	*Production, million tonne*	*% of Total Edible Oil*
Rape/mustered	1.82	36.21
Groundnut	1.22	24.30
Soybean	0.62	12.0
Coconut	0.44	8.80
Cotton seed	0.41	8.14
Sunflower	0.29	5.77
Sesame	0.16	3.18
Safflower	0.05	1.00
Linseed	0.03	0.60
Total	**5.04**	**100.00**

Source. Agriculture Statistics at a Glance, 2001.

Soybean production in the world during 2000 was about 162 million tonnes (Table 1.5). The five major soybean growing countries are USA (46 per cent), Brazil (20 per cent), Argentina (12 per cent), China (9 per cent) and India (3 per cent). The world soybean productivity is about 2.2 t/ha. It is expected that its production during 2003-04 would be about 203 million tonnes (Table 1.5).

Table 1.5: Estimated Production of Soybean in the World During 2003–04

Country	*Area, million hectare*	*Productivity, t/ha*	*Production, million tonne*	*% of Total Production*
USA	46	1.8	83	41
Brazil	19	2.8	53	26
Argentina	13	2.7	35	17
China	9	1.8	16	8
India	7	1.0	7	3
Others	6	1.5	9	5
Total	**100**	**2.03**	**203**	**100**

Comparative Cost of Protein from Different Sources

India produces about six million tonnes of soybean worth Rs 60,000 million as raw food/oil commodity. Its protein (2.4 million tonnes, 40 per cent in soybean) value in terms of dal (processed pulse) equivalent is 9.6 million tonnes (25 per cent protein) or 12.8 million tonnes of unprocessed pulses (75 per cent dal recovery or 18.75 per cent protein in unprocessed pulses). The production of pulses in India is about 15 million tonnes. It, therefore, shows that the contribution/share of soy-protein in comparison to traditional pulses is substantial, *i.e.*, 85 per cent. Processed soybean, in the form of full-fat soy-flour (FFSF), would cost Rs 20/kg in the retail market with 43 per cent protein and other nutrients whereas an average cost of one kilogram dal in the market is Rs 25 and its protein content is about 60 per cent of FFSF. Quality-wise, soy-protein is better than pulse-protein. In fact, soy-protein is of the best quality among all other plant proteins. Soy-based food item, like FFSF containing all nutrients as well as phyto-chemicals, is healthful and economically affordable by all sections of Indian population especially those living below the poverty line. Approximate cost of one-kilogram of protein from different sources is given in Table 1.6. Cost of one-kilogram soy-protein in the form of full-fat soy-flour is Rs 50 whereas the same in the form of *dal* is Rs 100. Protein efficiency ratio (PER) of soy-protein increases considerably when combined with cereal and legume proteins. Health benefits that could be derived from regular use of soybean in daily diet are given in Table 1.7.

Table 1.6: Approximate Cost of 1 Kilogram of Protein from Different Sources

Source	*Cost, Rs*
Full-fat soy-flour	50
Split pulses (dal)	100
Egg	250
Milk	300
Chicken	360
Fish	420
Meat	480

Source: Ali, 2000.

Table 1.7: Major Health Benefits from the Regular Use of Soybean in Daily Diet

Nutrient	*Health Benefit*
Protein	Lowers blood cholesterol
Carbohydrate	Relaxes constipation Good for diabetics
Fat	Prevents cardio-vascular diseases
Mineral	Overall health promotion
Vitamin	Overall health promotion
Phyto-chemicals	Prevent cancer, helpful in menopause and osteoporosis

Source: Danji, 2000.

Rationale for Promoting Soybean Foods in India

Soybean is one of the oldest food sources known to the man and it is recognized for its value in enhancing and protecting health. Soy-protein has all the eight essential amino-acids. The recent discovery of the value of soy-isoflavones and their role in disease prevention has created special interest of man in soybean, which has boundless food potential. It is well known that protein is essential for all and it is derived from plant and animal sources. Some of the potential food uses of soybean and its derivatives are given in Table 1.8.

Table 1.8: Potential Food-uses of Soybean and its Derivatives/Ingredients

Product/Ingredient	*Potential Food Applications*
Whole soybean	Soy-based dairy analogs, full-fat soy-flour, snack foods, sprouted bean, *tempeh, miso, natto* and sauce
Soybean oil	Cooking/shortening oil, salad oil, margarine, bread spread, and soylecithin
Soy-proteins (Edible soybean meal)	Texturised soy-protein, soy-protein concentrate, soy-protein isolates, defatted soy-flour, and dietary fibre
Soybean byproducts (hull and okara)	Single cell protein (SCP) and dietary fibre
Soybean crop residue (leaf, branches and stems)	Animal feed, fuel, and manure

Source: Ali, 2000.

In India, majority of the population (65-70 per cent) are vegetarian and get their dietary proteins from pulses and cereals and to some extent from oilseeds like soybean, groundnut and sesame. Proteins from animal sources like milk, meat, fish and poultry are costly and only a select group of population has access to it. About 40 per cent of Indian population is living below poverty line and does not have enough purchasing power to buy pulses, which are the major dietary protein sources for most of the Indians. In such a situation, use of soybean is one of the most appropriate options for India to augment its proteins supply at a price, which is affordable by the poorer sections of the population.

Soy-foods for Indian Population and Entrepreneurship Development

Indian snack foods like cookies, *sev, laddu, halwa*, roasted/fried soybean have been prepared by replacing 15-30 per cent of the traditional ingredients with soy-products and found acceptable by the people. Soy-fortified wheat flour gave good preparation of *chapatis, puri* and *paratha*. Chickpea-flour mixed with soy-flour resulted into a good quality like *sev* and *pakoda*. Soy-flour incorporated *idli/dosa* batter gave good and acceptable products. The consumers have found soy-cereal extruded snacks quite acceptable. Green soybean can also be used as vegetable. Soy-products can fit very well into various Indian food recipes provided it is properly processed and blended. Some of the soy-products, processing equipment and technology developed in India for food uses of soybean are given in Tables 1.9, 1.10 and Figure 1.1a and 1.1b. Technology and hardware is necessary to develop entrepreneurship in soy-food manufacturing and marketing.

Many small-scale entrepreneurs are interested in soy-based food manufacture. However, they wish to see the working of a complete plant to know the actual investment cost and the facts about plant capacity and product quality to judge the market potential to be sure of training facilities available for their production staff, before deciding to go for the establishment of a soy-food plant. Project reports along with economic analysis based actual commercial-scale plants are available at SPU, Central Institute of Agricultural Engineering (CIAE), Bhopal. The SPU can supply technology, prepare the project report, provide support services including consultancy for production and marketing of soy-products. Training in soy-based food production and machinery can also be arranged for individuals, groups and entrepreneurs for domestic use and commercial manufacturing.

Overcoming Barriers in Promoting Soy-foods

In spite of a number of good features of soybean, there are a few constraints associated with its food uses. These are beany flavour of soy-food products and oxidative instability of soybean oil. So far, two major approaches have been used to overcome these constraints. One is through innovative processing and the other is through plant breeding. Other means, such as marketing efforts, medical discovery, consumer education and change in dietary habits are also needed. Acceptance of soy-foods in India has been rather slow. However, now more and more people are consuming soy-food because of its economic, nutritional and health benefits.

The first and the foremost measure required is to create awareness among the masses about the economic, nutritional and health benefits of soybean and its products, using print and electronic

Soy-*paneer*

Full-fat Soy Flour

Okra Based Soy *gulab jamun*

Soy-muffins

Soy-nuts

Soy-yogurt

Figure 1.1a: Some of the Soy-foods Developed at SPU, CIAE, Bhopal (India)

Table 1.9: Soy-products and Processing Equipment Developed in India for Food Uses of Soybean

Soy-products	*Soybean Processing Equipment*	*Limited Scale Pilot Production Facilities*
1. Soy-*dal*	1. Grader	1. Full-fat soy-flour
2. Soy-flakes	2. Manual dehuller	2. Partially defatted soy-flour
3. Soy-flour	3. Power operated dehuller	3. Soy-milk
4. Soy-fortified biscuits (sweet)	4. Blancher	4. Soy-*paneer*
5. Soy-ice cream	5. Natural convection tray dryer	5. Soy-fortified bakery products
6. Soy-fortified biscuits (salty)	6. Forced convection tray dryer	6. Soy ice-cream
7. Soy-fortified bread	7. Multipurpose LSU type dryer	
8. Soy-fortified muffins	8. Three-roller flaking machine	
9. Soy-fortified bun	9. Two-stage roller mill	
10. Soy-snacks (roasted/fried)	10. Low-cost single screw forming extruder	
11. Soy-*amrakhand*	11. Plate type wet grinder	
12. Soy-milk	12. Modified oil expeller	
13. Soy-*paneer*	13. Low-cost steam generator	
14. Soy-yogurt	14. Soybean cake grinder	
15. *Tempeh*	15. Dough mixture	
16. Soy-*sattu*	16. Loaf volumeter	
17. Okara based *burfi* and *Gulabjamun*	17. Lever-type *paneer* pressing device	
18. Soy-sauce	18. Screw-type *paneer* pressing device	
19. Soy-*srikhand*	19. Cottage level soy-*paneer* plant	
	20. Biscuit-cutter/ moulder	

Source : Ali, 1996 (b), Ali, 2000 and SPU, CIAE, Bhopal.

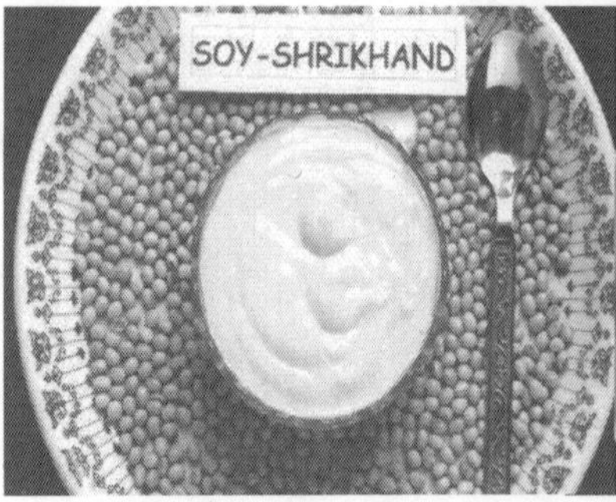

Soy-*shrikhand*

Soy-*amrakhand*

Soy-biscuits

Soy-*sattu*

Cottage Level Soy-milk Plant

Making and Marketing of Soy-based Products at Ludhiana

Training on Soy-products at SPU

Figure 1.1b: Some of the Soy-foods and Machines Developed at SPU, CIAE, Bhopal (India)

media. The second measure is training in manufacture and marketing of soy-based food products and machinery for individuals, groups and entrepreneurs. The third measure is to provide technical support to the potential entrepreneurs through project report, consultancy and service support. Fourth measure is the strong political will and positive government policies to encourage production and utilization of soybean through fiscal incentives like soft-loan and tax-rebate/concessions. Fifth is the research and industry linkage to refine the product and modify the technology with time for better efficiency and output. For all these to happen, there must be a commitment and missionary zeal by all those involved in soybean production and the value-chain.

Table 1.10: Traditional and Non-traditional Soy-products being Used in India

Traditional Products		*Non-traditional Products*
Non-fermented	*Fermented*	
1. Cooking oil	1. Yogurt	1. Texturised soy-protein
2. Hydrogenated fat	2. *Tempeh*	2. Health foods
3. Flours, flakes and grits	3. Sauce	3. Lecithin
4. Roasted/fried nuts		4. Protein concentrate
5. Sprouted beans		5. Protein isolate
6. Cooked beans		6. Protein hydrolysate
7. Milk		7. Liquid protein
8. *Paneer* (Tofu)		
9. Bakery products		
10. Splits (*dal*)		
11. *Vadi*		

Soy-food Entrepreneurs

The primary interest in soybean in India has been the oil. However, increasing attention is now being paid to the protein potential that soybean offers. As of now, about 15 per cent of total soybean production in India goes for direct food and feed uses, 10 per cent for seed and 75 per cent is processed for oil and protein. Crop residues like leaves and fine straw are used as feed and the hard and woody stem for fuel in the rural areas.

The major food uses of soybean in India are of its oil, protein and lecithin. The present soybean oil processing capacity is about 15 million tonnes of soybean and that for food is about one million tonne. The Indian soybean industry has the capability to process soybean for food, feed, pharmaceutical and industrial applications. The present soy-food industry has about 330 units which may be grouped under different categories, as under:

1. Oil extraction plants: 154
2. Food manufacturing units: 125
3. Equipment manufacturers: 30
4. Trading houses: 15
5. Government and other agencies: 06

Under soy-food entrepreneurship development programme, the Soybean Processing and Utilization (SPU) Centre, Central Institute of Agricultural Engineering, Bhopal till June, 2003 has trained 432 potential entrepreneurs in soy-milk and soy-*paneer* production and marketing (Table 1.11). The SPU has also trained 386 rural/urban women in processing and utilization of soybean at domestic level (Table 1.12) and also to establish micro-enterprise by the women. Out of these, 90 entrepreneurs have already started their enterprises. There are six manufacturers/suppliers of cottage level soy-*paneer* plant in India. Out of 90 new entrepreneurs, six most successful entrepreneurs are located in Ludhiana, Jabalpur, Ambala and Calcutta.

Table 1.11: Potential Entrepreneurs Trained in Soy-based Dairy Analogs at SPU, CIAE, Bhopal (December 1995 to June 2003)

Year	*Participants*
1995-96	21
1996-97	47
1997-98	56
1998-99	47
1999-2000	42
2000-2001	71
2001-2002	54
2002-2003	75
2003-2004 (up to June 2003)	19
Total	**432**

Table 1.12: Women trained at SPU, CIAE, Bhopal (1998-99 to 2003-04)

Year	*Participants*
1998-99	30
1999-2000	35
2000-2001	93
2001-2002	112
2002-2003	85
2003-2004 (up to May 2003)	31
Total	**386**

There are three manufacturers of soy-fortified biscuits and five of soy-fortified wheat flour. Edible oils and texturised soy-proteins (TSP) are being manufactured by a number of enterprises. Roasted/fried soybean as snack food is also being manufactured and marketed. Cottage/small scale soy-based food industry is increasing in number, every year. It has been reported that there are many tiny/micro soy-based food enterprises but it is difficult to list them. However, SPU at Bhopal is trying to collect such information and catalogue it for further use. The action plan for soy-food promotion in India is shown in Table 1.13.

It may be construed that the major food uses of soybean, as of now, in India are edible oils, texturised soy-proteins (TSP), flours, bakery products, milk/*paneer* (tofu), soy-protein concentrate/isolates/hydrolysates, lecithin and others. The need is to create awareness about soy-products and their benefits and make available such products in the market through small scale decentralized soy-food processing enterprises. Domestic level processing and utilization of soybean for food and feed need to be given priority especially in the rural sector. The centre and state developmental agencies and NGOs may come forward and make the implementation plan. The hardware and technology for a number of soy-foods matching with the Indian food recipes and food habits are available at SPU, CIAE, Bhopal and the Centre has the infrastructures and specialists to impart training for human resource and entrepreneurship development in soy-foods. The need is to make use of such indigenous facility for the benefits of the Indian people.

Table 1.13: Action Plan for Promotion of Soy-foods in India

Short-term (1-3 Years)	*Long-term (3-5 Years)*
Awareness creation and generic promotion of soy-foods	Generic promotion of soy-foods through print and electronic media
Training of rural/urban individuals and/or groups at domestic level soybean processing for food uses	Development of soybean varieties with specific end uses (Food/oil/ pharmaceutical)
Human resource and entrepreneurship developments in soy-foods	Application of recent food processing technology such as extrusion cooking, high pressure membrane separation, enzyme processing, SCF extraction, microwave heating, etc. to have newer and diversified soy-products for food and feed.
Preparation and notification of quality standards for soy-foods	Development of soy-based specialty foods
Establishment of soy-food enterprises for manufacturing and marketing of soy-foods	Formation of Soy-foods Manufacturers Association of India to take care of large scale soy-food manufacturing and marketing in India and abroad.

Soy-foods enhance human health at an affordable cost. Hence the need is to promote soy-foods for all and by all means

References

Ali, Nawab.1996a. Soybean Processing and Utilization in India. The Country Report presented in the Regional Expert Consultation on Soybean Processing and Utilization in Asia. FAO Publication. 1996/2, Bangkok, Thailand.

Ali, Nawab. 1996b. The Indian Soybean Industry and present status of the emerging trends. The Indian Food Industry. 15(3)/1996. May/June.

Ali, Nawab. 2000. Soybean processing technology and promotional efforts for its food uses in India. Proceedings of the 3rd International Soybean Processing and Utilization Conference (ISPUC-III) held at Tsukuba, Ibaraki, Japan. October 15-20, Pages 327-328.

Anonymous. 2001. Agricultural Statistics at a Glance. Department of Agriculture and Cooperation, Ministry of Agriculture, Government of India, New Delhi.

ASA. 1999. American Soybean Association, New Delhi.

Danji, Fukushima. 2000. Recent progress in research and technology for processing and utilization of soybeans. Proceedings of ISPUC-III held at Tsukuba, Japan. October 15-20. Pages. 11-16.

Gopalan, C, BV Rama Sastri and SC Balasubramanian, 1974. Nutritive Value of Indian Food, National Institute of Nutrition, Hyderabad, India.

Singhal, V. 2003. Indian Agriculture–2003. Indian Economic Data Research Centre, Vikaspuri, New Delhi.

Soy-update. 2004. Monthly News Magazine for Soy-foods and Feed Industry. Vol.-1(10):17. October, 2004.

Chapter 2

Postharvest Management and Value-Addition: Prospects and Opportunities

S.M. Iilyas and R.K. Goyal

Central Institute of Postharvest Engineering and Technology, PAU Campus, Ludhiana – 141 004 (Punjab)

The dynamic, non-liner system, easily affected by the factors beyond anybody's control, our agriculture has received many accolades but is also increasingly criticized for its sluggish growth and declining contribution to the national Gross Domestic Production (GDP), at present 27 per cent but going down. However, no sector of economy has been able to dislodge it from its exalted status of being its backbone–the declining clout not with standing. Having said this, it is also to be recognized that agriculture is fast losing its shine and urgent re-look at our priorities is over-due. Indian agriculture, which has hitherto been a way of life has to transform itself into a new *avatar* of commercial agriculture and do what it takes to become such. Many models have been suggested from brilliant to hare-brained for achieving this breakthrough. However, any model, which talks about complete break from the past, may not succeed in our milieu because in a country of our size it takes more than a single factor, how so ever important, to change the scenario. There will have to be many paradigm shifts. To start with, we must come out from the blame-game and go ahead with whichever model of change we are comfortable with.

It has been observed that our agriculture and allied sector continue to be geared up for raw material supply and not much beyond. Selling raw material is a sign of primitive economy and agriculture is the biggest follower of this practice. The return on investment from sale of raw material can never reach the level of return from sale of value-added produce for which any number of examples can be cited. One example would nevertheless, be highly illustrative. In the glut season, potato sells for less than Rs. 2-3/kg, but the branded potato chips sell for Rs. 250-300 per kg, both the farmers and consumers getting a raw deal. This happens because of many natural and man-made reasons, but the main reason is that there are so many intermediaries between the producer and the consumer which need to be brought down to more manageable level (as it happens in the developed countries). The farm gate price in India as a percentage of the consumer retail price is quite low and needs to be raised. The less remunerative agriculture is a cause of great worry and efforts need to be made to convert it into highly remunerative one as it will ensure greater infusion of money (in the form of inputs etc.) and technology forming a virtuous circle. The postharvest management involves reduction of huge losses, safe handling, transport, storage and most importantly processing and value-addition. This will ensure more money into the pocket of farmers and also availability of high quality raw and processed product at competitive price to the consumers (domestic as well as foreign).

Indian Agriculture

Indian agriculture has made great strides, which has been often praised but seldom replicated. The country with a production of 750 million tonnes of food commodities is only second to China with 830 million tonnes and there is no doubt that we shall overtake it in the next a few years. India's status in world agriculture and share of raw materials is given in Tables 2.1 and 2.2. While that sounds good and heartening, it is also a fact that out of this 750 million tonnes, the country may actually be getting around 600 million tonnes only, thus roughly losing around 764 billion rupees every year as shown in Table 2.3.

Table 2.1: India's Position in World Agriculture

Item	*India*	*World*	*India's*		*Next to*
			Share, %	*Rank*	
			1. Area (million ha.)**		
Total area	329	13389	2.4	Seventh	Canada, U.S.A., China, Brazil, Australia, Russian Fed.
Land area	297	13048	2.3	Seventh	U.S.A., China, Canada, Brazil, Australia, Russian Fed.
Arable land	163	1382	11.8	Second	U.S.A
Irrigated area	57	242	23.6	First	
			2. Population (million)		
Total	960	5849	16.4	Second	China
Agriculture	256	1302	19.3	Second	China
			3. Economically active population (million)		
Total	420	2827	14.9	Second	China
Agriculture	256	1302	19.3	Second	China

Contd...

Table 2.1–Contd...

Item	India	World	India's		Next to
			Share, %	Rank	
			4. Crop-production (million tonnes)		
Cereals	200	2096	10.6	Third	China, USA
Wheat	69*	610	11.3	Second	China
Rice	80	573	21.5	Second	China
Coarse grains	31	914	3.5	Fifth	USA, China, Brazil, Russian-Fed
Potatoes	19	295	6.5	Sixth	China, Russian-Fed, Poland, USA, Ukraine
Total pulses	15	58	25.9	First	
Ground Nut	8*	30	26.7	Second	China
Tobacco leaves	0.56 #	8.24	6.8	Third	China, U.S.A.
Rapeseed	7	35	20.0	Second	China
Coffee (green)	0.21*	5.56	3.8	Ninth	Brazil, Colombia, Indonesia, Mexico, Ethopia, Uganda, Vietnam, Gautemala
Sugarcane	265	1241	21.4	Second	Brazil
Tea	0.79 #	2.73	28.9	First	
Jute and Allied Fibres	1.72 #	3.27	52.6	First	
Cotton (lint)	286*	19.7	14.5	Third	USA, China

Contd...

Table 2.1–Contd...

Item	*India*	*World*	*India's*		*Next to*
			Share, %	*Rank*	
5. Livestock numbers (million head)					
Cattle	209	1334	15.7	First	
Buffaloes	92*	167	55.1	First	
Camels	1.52 #	1950	7.8	Third	Somalia, Sudan
Sheeps	56*	1064	5.3	Third	Australia, China
Goats	121*	703	17.2	Second	China
Chicken	343*	13413	2.6	Seventh	China, USA, Brazil, Russian-Fed, Mexico, Indonesia
6. Implements ('000 Numbers)**					
Tractors-in-use	1400 #	26345	5.3	Fourth §	USA, Japan, Italy
7. Vegetables, Fruits, Milk and Eggs (million tonnes)					
Vegetables (including Melon)	54.97 #	595.56	9.2	Second	China
Fruits (excluding Melon)	37.13 #	429.45	8.6	Third	China, Brazil
Milk	80.00	546.64	13.2	First	
Eggs	1.61*	51.75	3.1	Fifth	China, U.S.A., Japan, Russian-Fed

Note: # FAO Estimate, § Largest producer of tractors in the world, * Un-official figure and ** Figures of 1996.

Sources: 1. FAO Production Year Book 1997 and 2. Agricultural Statistics at a Glance, 2000.

Table 2.2: India's Share of Raw-materials in the World

Raw-material	*Share, per cent*
Fruits and vegetables	
Mangoes	52
Bananas	18
Lemons	19
Dry onions	11
Cauliflowers	38
Cereals and pulses	
Wheat	11
Rice	21
Pulses	27
Dry beans	22
Chick peas	67
Lentils	28
Oilseeds	
Groundnut	27
Castor seeds	70
Rapeseed	20
Nuts	
Coconuts	20
Cashew nuts	21
Milk and its products	
Buffalo milk	62
Goat milk	20
Butter and *ghee*	21
Evaporated milk	10
Plantation crops	
Tea	27
Sugarcane	22

Source: Ministry of FPI, Govt. of India, New Delhi.

Present Status

India with 16 per cent of world's population, 4.5 per cent water and 2 per cent land area is the second largest producer of food (next to China), which includes 200 million tonnes of food grains and about 132 million tonnes of fruits and vegetables. Due to the multitude of postharvest factors, the postharvest loss in fruits and vegetables is four to five times higher than in food grains. The wastage-cost of fruits and vegetables alone is estimated to be 300 billion rupees each year. While the production scenario is heart warming, hardly 2 per cent of the horticultural produce goes for commercial processing, whereas this figure is more than 75-80 per cent in developed countries. Due to improper postharvest management and lack of adequate processing facilities, a huge quantity of nutritive food materials are lost in our country. In the horticultural sector, the situation is worse since it is often found that there is glut of a particular horticultural commodity in one part of the country when it is scarce in other parts. Prices of seasonal fruits and vegetables fluctuate greatly and during the period of maximum availability, the prices become un-remunerative to the farmer. At other times, these commodities are so highly priced that the ordinary consumer finds it beyond his purchasing power. It is unfortunate that unscrupulous traders and middlemen (5-7 in India as compared to 2-3 in developed countries) get away with most of the profit leaving the producer high and dry. It is reckoned that the farmer/producer gets only 25-30 per cent of the price paid by the consumer.

Table 2.3: Level of Production of Food Commodities and Estimated Postharvest Losses

Type of Food Commodity	*Present Level of Production*			*Postharvest Losses*	
	Quantity, Million Tonne	*Average Price, Rupees/Tonne*	*Value, Billion Rupees*	*Quantity Million Tonne*	*Monetary Value, Billion Ruppes*
Durable (cereals, pulses and oil-seeds)	215	10000	2150	21.5	21500
Semi-perishables (potato, onion, sweet potato, tapioca)	40	3000	120	8.0	2400
Perishables (fruits, vegetables, milk, meat, fish and eggs)	140	15000	2100	35.0	52500
Total	**395**	-	**4370**	**64.5**	**76400**

Source: Pandya and Ali, 2000.

Scope

The avoidable wastage of valuable commodities can be checked if these are processed into value-added products or adequately distributed in different parts of the country. There is a great scope of both domestic and export trade in our country by improving the postharvest distribution and processing facility through an organized yet decentralized system. If the fresh and processed fruits and vegetables are evenly marketed from the places of abundance to the place of scarcity, not only will the consumers get the produce at a reasonable price but also the producer will not be forced to sell at throw-away prices. Some of the techniques are elaborated which are generally not followed in our country and that can be of some practical use to achieve our national objectives.

To attract and retain youth in agriculture, it is essential that agriculture becomes both intellectually stimulating and economically rewarding. For this, especially in the context of resource poor farmers and women, it is important that the economic benefits of agro-processing and agri-business are taken to the rural areas. Most agriculture and food processing industries around the world operate in a manner that would ensure economies of the scale. This also leads to economization of farming and processing. Incremental job opportunities in the agri-food sector are therefore, limited, especially in advanced countries. Developing countries such as India are somewhat differently placed. It is reckoned that this sector will provide maximum opportunities for income and employment generation.

National Food Processing Policy

The Government has come out with a draft national food processing policy with a vision to motivate farmers and food processors and to provide interactive coupling between technology, economy, environment and society for steady development of food processing activities to build up a substantial base for production of value-added agro-food products for domestic and export markets with a strong emphasis on food safety and quality enabling the farmers especially to realize direct benefits of new technology and marketing network and to ensure adequate availability of quality food products for the consumers at economic prices.

The policy seeks to create an appropriate environment for the entrepreneurs to set up food processing industries through:

1. Creating enabling environment
2. Infrastructural development
3. Backward linkage
4. Forward linkage

Food Processing Industry

Food Processing Industry is of enormous significance for India's development because of the vital linkages and synergies that it promotes between the two pillars of the economy, namely industry and agriculture. It will give India the potential to become number one in food production with sustained efforts. The growth potential of this sector is enormous and it is expected that the food production will double in the next 10 years, marking improvement in consumption of value-added foods.

Growth of the food processing industry will also help in bringing immense benefits to the economy, raising agricultural yields, meeting productivity, creating employment and raising the standard of very large number of people through the country, specially in the rural areas. Economic liberalization and rising consumer prosperity is opening up new opportunities for diversification in food processing sector. Liberalization of world trade will further open up new vistas for growth. Food processing has the largest employment generation potential. It generates employment for 54,000 persons per 10 billion rupees investment, whereas in textiles and paper industries it is 45,000 and 25,000 persons respectively. Altogether 1.27 million job opportunities are likely to arise in the food-processing sector with the implementation of the approvals given and industrial entrepreneurial memoranda filed so far.

Although we are one of the largest producers of raw materials for the food processing industry in the world, the industry itself is extremely underdeveloped in India. Less than 2 per cent of fruit and vegeTable production is processed as compared with 30 per cent in Thailand, 70 in Brazil, 78 in Philippines and 80 in Malaysia. The value-addition in the food sector is still very low at 7 per cent.

Agro-processing

Primary Processing of Cereals, Pulses and Oilseeds

Cereals

Milling of cereal used to be done by women mostly using manual and foot-operated chakkies for their household consumption and occasionally to help the relatives. Now the milling has been replaced by *chakkies* for making wheat flour and dehulling/shelling of rice. The pulse processing has entirely gone into hands of *dhal* mills.

The flour *chakkis* are of 5-10 hp and produce wheat flour of different fineness. Mini burr mill developed at Central Institute of Agricultural Engineering (CIAE), Bhopal (capacity 10-75 kg/h, 0.5 hp electric motor) have been found as versatile size-reduction equipment for the production of flour, *dhal, besan* and powered spices (chili, coriander, turmeric etc.). Central Food Technological Research Institute, Mysore has developed grain mill of the capacity of 100 kg/h and it produces flour (*atta*), refined flour (*maida*) and grits (*suji, dalia*).

Rice milling which was earlier done by hand/foot pounding has been replaced by paddy hullers/shellers. There are a large number of rice mills of varying capacity in the country. Parboiled rice is prepared in certain parts of the country. Low cost methods, in addition to commercial large-scale methods, have been developed for parboiling. Improved versions of machines for use at small scale or as cottage industry have been developed and are commercially available. Women can easily operate these equipments.

The primary grain processing, particularly processing and milling of rice and wheat has substantial capacity in the country. In the last five years, there has been a marginal increase in the number of flour mills adding to the capacity. Table 2.4 gives an idea about the large number of milling equipment in India.

India is the largest exporter of Bastmati rice and due to high competitiveness, the rice mills have installed most modern milling equipment like colour sorters etc. Solvent extraction milling is also quite promising.

Pulses

About 75 per cent of the pulses produced in India are milled into *dhal*. Traditionally *dhal* was produced after simple pre-treatment with the help of lightweight stones (grinder/*chakki*) which leave some husk (5-10 per cent) having low market acceptability. Now simple *dhal* mills have been developed at many research and development institutions and can be easily operated for *dhal* production at cottage level.

Table 2.4: Rice, *Dal* and Wheat Milling Equipment

Type	*Number*
Hullers	90,000
Shellers	4,500
Huller-cum-sheller	8,300
Modern rice mills	33,000
Dal mills	12,000
Roller mills	750
Flour mills	2,66,000

The main objective of pulse processing is to remove hulls as cleanly as possible. This is accomplished by giving treatment (mostly hydro-thermal) for weakening of bond. In the traditional mills, water is sprayed by hand on the pulses and then sun-dried or in some cases dried in mechanical dryers. In further improved mills, a flight mixer does water application. After water application, tempering is required to get uniform effect as well as curing action. In the modern method, pitting (cracking of hull) is quite popular for facilitating water penetration. Pitting also helps in crushing shriveled grains and mud particles of the size of grain which could not be removed during cleaning and grading.

Traditional practice of *dhal* milling is time-consuming, more labour intensive and there is much loss of material with average milling yield of 65-74 per cent only as against 80-84 per cent which can be achieved. The milling of pulses on a commercial scale is generally based on dry-processing technique, where the major unit operations of de-husking and splitting are mechanized. Dehusking and splitting are either done in a single operation or more advantageously, as independent operations. There are presently about 700 *dhal* mills processing different pulses in the country. Types of mills, capacity and kind of pulses processed are indicated in Table 2.5. Significantly higher yields may be obtained when legumes

Table 2.5: Status of *Dal* Milling Industry in the Country

State	*Type of Mill*	*Capacity, tonne per day*	*Type of Machinery*	*Type of Pulse*
Andhra Pradesh	Large	10 to 20	Graders, rollers, vertical *chakkies* (plate mill)	Mung-bean, urd-bean pigeon-pea
	Small	1.5 to 3.0	Vertical *chakkies*, roller, aspirator	
Bihar	Large	15	Graders, rollers, vertical *chakkies*	Lentil, *khesari*
Gujarat	Large	20 to 50	Smooth and rough emery rollers, size graders, impact splitters	Pigeon-pea
	Small	1.5 to 2.0	Emery rollers and stone *chakkies*	
Haryana	Large	30 to 50	Size grader, rollers (pitting) under-runner disc sheller and mixer-cum- polisher	Pigeon-pea, chick-pea, peas
	Small	0.2 to 0.5	Vertical stone mill (flour mills) or paddy hullers	
Madhya Pradesh	Small	1.0 to 8.0	Agra type *dal* mill toller, size grader, mixer, vertical plate mill and polisher	Chick-pea, urd-bean, (only small quantity) Indore zone: pigeon-pea, chick-pea, mung-bean, urd-bean. Hoshangabad region: pigeon-pea, chick-pea

Contd...

Table 2.5–Contd...

State	*Type of Mill*	*Capacity, tonne per day*	*Type of Machinery*	*Type of Pulse*
Maharashtra	Large	20–40	Modernized (rollers and *chakkies*) mechanical driers	All pulses
	Small	0.2 -0.3	Combined emery rollers and horizontal *chakkies*. CFTRI designed pulse dehusker	
Orissa	Only small mills	1.0 to 1.5	Emery roller grader and polisher	Mung-bean, urd-bean, horse-gram
Rajasthan	Large Small	10 1.5 to 2.0	Disc sellers (splitting), emery rollers (dehusking), reel sieves (grading)	Chick-pea, mung-bean, moth-bean
Tamil Nadu	Large	5 to 10	Emery rollers (2 to 3) reel sieves, water mixer and polishers	Urd-bean, pigeon-pea, mung-bean, horse-gram
	Small	0.5 to 2.0	Plate mill/roller grader and aspirator	
Uttar Pradesh	Large	2.0 to 5.0	Agra-type machinery, under-runner disc sheller, vertical plate mill tapered rolls	chick-pea, pea, mung-bean, urd-bean (splits), pigeon-pea, lentil
	Small	2.0 to 4.0	Units from Kanpur vertical plate mills/burr mill	
West Bengal	Large and Small	0.2 to 5.0	Emery coated rollers either tapered or cylindrical cleaning and grading units and polishers	Urd-bean, lentil, peas, chick-pea, *khesari*

Source: Narsimha *et al.*, 2000.

are milled by improved method as compared to traditional methods as can be seen from Table 2.6.

Table 2.6: Average Yield of *Dal*

Sl.No.	*Legume*	*Kernel Content After Removing Husk, %*	*Head Dal Yield by Improved Method, %*	*Yield by Traditional Method, %*
1.	Chick pea			
	(a) Big grains	88.5	84	74
	(b) Small grains	87.5	83	72
2.	Pigeon pea			
	(a) Big grains	88.5	85	75
	(b) Small grains	85	82	68
3.	Mung gram	89.5	84	65
4.	Black gram	87.5	83	71
5.	Lentil	88.5	78	76
6.	Lathyrus bean	88	83	76
7.	Cow pea	89.5	84	—
8.	Kidney bean	90.3	86	—
9.	Horse gram	87.5	81	—
10.	Soy-bean	89.5	86	—

Source: Kurien, 1977.

Dhal milling units are notorious for their noise and dust pollution, which is required to be reduced/removed. Further, energy use is also poor and needs to be avoided and better power transmission mechanism required. For pre-treatment, work on the use of enzymes is encouraging but much more is required.

Oilseeds

India has a multi-tier extraction involving traditional *kolhus/ ghanis* located in villages, which leave 10-15 per cent oil with the cake followed by mechanical oil expellers located in semi-urban and urban areas which may leave 5-7 per cent oil with cake. In oilseed processing, dehulling/decortication is an important unit operation. A number of such units have been developed and are available in the market for various oilseeds. Women can easily operate groundnut decorticators, sunflower decorticators, castor shellers, soybean dehullers etc. For oil extraction, they can use power-driven portable power *ghanies* and Table oil expellers, which are commercially available. There are about 3,00,000 mechanical oil expellers (2,50,00 *ghanis* and 65,000 expellers) and 600 solvent extraction plants producing oil worth 10 billion rupees. Considerable work has been done on single and double stage screw press, steam cookers, disintegrators, filtering gadgets/processes, solvent extraction equipment and refining for processing of oilseeds. Work has been done on physical, chemical and enzymatic pre-treatments of oilseeds and results have been encouraging.

Solvent extraction of oil from oilseeds by using solvents (generally hexane) is becoming very popular worldwide as it leaves less than 1 per cent oil in the oilcakes. There has been wide concern against the removal of traces of solvent from the cake and as a result safer solvents like isopropyl alcohol (IPA) are being preferred. Continuous counter current percolation system is however more popular because of its better efficiency. Refining follows oil extraction, which is achieved through degumming, neutralizing, bleaching, winterizing and deodorization. Use of oilcake has been mainly as cattle feed. Even if half of the estimated cake/meal in India (about 5-6 million tonnes/year) is upgraded and utilized for human consumption, it would add 2-3 million tonnes food and even a million tonnes of protein for diet. Non-traditional oilseeds (1.1 million tonnes annually) are quite important for Indian economy especially for the livelihood of tribals. If these oil sources are exploited in a systematic way, sizeable quantity of edible/non-edible oils could be added to the oil grid and it can provide employment for 500 million man-days or 1.6 million workers per annum.

Recent advance in oil expelling is supercritical CO_2 extraction with or without a co-solvent (ethanol or IPA) at very high pressure and high temperature, which has shown promise. Supercritical fluid (SCF) has versatile extraction capability, which comes from the innate solvent power that originates from the molecular association of the fluid in its supercritical domain. Once the liquid molecules are placed at a temperature and pressure higher than their critical point, the inter-molecular distances and forces can be closely controlled by independently controlling either pressure or temperature. This control is exploited in supercritical fluid extraction.

Postharvest Management, Processing and Preservation of Fruits and Vegetables

Most of processed fruits and vegetables (juice and pulp, jams and jellies, beverages, *purees*, ketchups, squashes, concentrates, pickles, resins etc.) are prepared in cottage or small scale sector on account of round the year availability for a wide variety of fruits and vegetables as well as due to large scale domestic level processing done by women. Defective, small size products left after grading for Table uses can be easily processed to a great economic advantage. New techniques for extending shelf-life of fruits have been developed and commercialized like waxing, packaging in perforated polybags, cold modified/controlled atmosphere storage etc. Presently postharvest operations for commercial/organized marketing are undertaken at the urban or consumer level. If undertaken at production catchments level rural women can be encouraged for sorting, cleaning, washing, waxing, packaging and also minimal/ primary processing.

Pre-harvest spraying of fungicides on kinnow plants at 45, 30 and 15 days before harvest with 0.2 per cent Biltox-50 is better in reducing fruit-drop followed by 0.3 per cent Dithane M-45 and 0.2 per cent Bavistin. It has also been reported that Blitox-50+ 1 per cent Calcium nitrate or 0.3 per cent of Bavistin was sprayed at 15 days before harvest to minimize fruit-rot in un-waxed fruits during storage under cold-store condition. Although, it is a higher dose hence 0.2 per cent concentrated Bavistin either with calcium-nitrate or alone may be applied before harvest. Wax application reduced fruit-rot as well as weight-loss during storage. Total 10 fungi were identified from the surface of kinnow fruit and out of those Penicillium digitatum, P. italicum and A. citri were found most prominent fungi to cause rotting. Postharvest treatment of fruits with 1000 ppm concentration of Bavistin was significantly effective against P. iltalicum in un-punctured fruits which could be stored safely upto 60 days under cool chamber. Fruits treated with fungicide in combination with wax-emulsion (Zividar+Bavistin 0.1 per cent) and stay-fresh (shine) + Bavistin 0.1 per cent were found better than other treatments. Fruits treated with hot water immersion at 50° C for 2 minutes did not rot even up to 90 days of storage at 5 ± 1° C. It was also observed that if fruit got injury then no treatment except D.

hansenii is able to check the growth of *P. italicum* on the its surface. D. hansenii + Mustard oil in 50:50 ratio reduced fruit-rot as well as further growth of Penicillium spp. on the surface of the fruit. Corrugated fibre board boxes (7-ply) were found suitable for packaging of kinnow fruit and it reduced the transportation and subsequent storage loses as compared to other packaging materials like wooden boxes, bamboo baskets and gunny bags.

Tomatoes harvested at breaker-stage and carried in plastic crates with grass as the cushioning material account for the least losses (4 per cent for 400 km distance of transportation). In other packaging and cushioning materials, losses are 0.2 to 0.5 per cent higher than that for breaker-stage and 5.0 to 6.0 per cent higher at pink-stage. Tomatoes harvested at breaker-stage using straw as the cushion material during transportation cause 1.1 per cent losses for 100 km distance of transportation in plastic crates. In other materials, losses are 0.6 to 0.7 per cent higher for breaker-stage and 2.3 to 4.8 per cent for pink-stage.

The postharvest losses of tomatoes at different levels are given in Table 2.7.

Table 2.7: Different Type of Losses at Various Levels

Types of Losses	*Range of Losses*
Losses at farm level	2-10 per cent
Losses at wholesale level	Monetary loss of 30-40 per cent price (5-8 per cent tomatoes are sold at 30-40 per cent reduced price).
Losses at retailer level	Physical loss 2-3 per cent (Monetary loss of 70-80 per cent of price in about 10 per cent of tomatoes)
Losses at cold store level	6-8 per cent
Losses at processing industry	Loss of 2 kg of tomato pulp for preparing each 1 kg of ketchup of TSS 26–28 °Brix

With proper postharvest management practices, which include right stage of harvest with proper harvesting technique followed with sorting, packaging and transporting as per the recommendations described above, better quality of tomatoes with the least postharvest losses can be ensured.

Jaggery and Khandsari Preparation

Jaggery (Gur) is a traditional sweetner and snack food in India which is nutritionally superior to refined sugar. Quality jaggery is in demand not only in the domestic market but for export also. About 45 per cent of sugarcane produced is utilized for production of jaggery and khandsari which meets more than 42 per cent of total sweetner demand in the country. Juice extraction, concentration and moulding into different shapes and sizes at processing centres and preparation of various items like cookies etc. is generally done by women. Technology has also been perfected for the preparation of solid (cubes), liquid or powdered jaggery, chocolates and toffees etc.

Milk and Milk Products Processing

India produces 80 million tonnes of milk annually and only 15 per cent (the highest percentage for processing of food-produce) is being processed. There is a large scope for processing of milk from chilling to the preparation of the following products:

1. Fluid milk: boiled, chilled, sweetened, flavoured
2. Concentrated products like *kheer* and *rabdi*
3. Fermented products like curd, yoghurt, *lassi, shrikhand*
4. Coagulated products: *paneer, chhena*
5. Frozen deserts: ice cream, *kulfi*
6. Fat-rich products: butter and *ghee*
7. Sweets/confectionaries: *burfi, peda, gulab jamun, rasgulla,* milk cake etc.

Simple, dairy equipment are available for preparation of cream, butter and ghee to replace the traditional churning methodology. Women can easily operate these.

Preparation of Novelty Food Items, Snacks and Bakery Products

Snack-foods are largely popular in India and have a very large variety. It is estimated that snack-foods trade runs into millions of rupees every year. Besides the huge demand for domestic consumption, these items especially *namkeen, papads*, flaked cereals, roasted breakfast cereals, puffed grain, roasted and fried legumes and nuts, extruded/formed snacks like *'sewe' 'murukku*, banana/

jackfruit/potato/cottage/sago chips, a wide variety of sweet meats are also prepared at a small scale. The women are mainly involved in the preparation of these items. In the last a few years instant noodles, soft drink concentrates, powders, *idli, vada, gulab jamun* and *jalebi* mixes for sweets and savouries have been developed. Technologies have been developed for adoption at the cottage level. Preparation of full-fat soy-flour, soy-biscuits, soy-milk, soy-*paneer*, soy-yoghurt (curd) and soy-ice cream is easy and can be undertaken by women on a commercial scale.

Value-addition

The value-addition of farm-produce could be undertaken due to different factors including increase in the price of the produce, sale before or after the glut and higher market demand etc. The farmer must have means and capacity to hold the farm produce for longer periods for which he requires an appropriate storage structure and the desired technology to reduce storage losses and transform the produce into a product which fetches higher price, is able to hold little longer and is available out of season etc. Thus processing is a major value-addition activity to the farm produce. The processing technology depends upon the type of raw-produce and is location and crop-specific. Various postharvest operations include cleaning, grading, drying, dehydration, dehulling, decortication, shelling, peeling, grinding/milling, polishing, extraction, curing, treatment, storage, packaging, transportation etc. The postharvest losses of cereals, pulses and oilseeds are about 10-12 per cent which are quite large. Even if these losses can be reduced by half, it will make a difference to the economy of the country. Table 2.8 quantifies the benefits through value-addition.

Table 2.8: Estimated Value-addition to Agricultural Produce (1995-96)

Level or Processing	*Value-addition, Billion Rupees*
Primary processing	1080
Secondary and tertiary processing	220
Products produced/processed in organised sector	100
Products/processed in un-organised sector	1200

Source: Singh, G. 1997.

Causes for Low Level Processing

1. Most Indians prefer to consume fresh fruits and vegetables for a variety of reasons mainly due to prevalence of warm sub-tropical climate where the chances of spoilage are high.
2. High cost of processed foods which is beyond the reach of the poor people. Enhanced prices may be due to low turnover, low productivity, seasonality of produce and high extent of spoilage.
3. Poor quality of raw material which affects the quality of processed goods.
4. A plethora of rules and regulations resulting in long processing time, which inhibit a budding entrepreneur. Many of the laws were framed some 50 years ago and need revision.
5. General lack of knowledge about the processes, equipment and technology due to lack of availability of information from reliable sources. There is a need for developing a strong data base.
6. Fewer opportunities for human resource development resulting in shortage of trained manpower.
7. Lack of financial support on soft terms and absence of incentives like tax exemption, relaxation/rebate etc. for starting a new unit. It may be noted that in India, tax on processed food is amongst the highest in the world. A tax holiday for a period of 10 years would be a good idea.
8. Lack of infrastructure including electricity, good quality water, communication facilities etc. makes it very difficult to sustain processing units. The power blackouts for hours/days are not unknown in many parts of India.
9. Inadequate market intelligence due to lack of infrastructure creates problems for marketing of goods to fetch better price. Market intelligence network needs to be established.

Factors Attracting Farmers for Agro-processing Enterprise

1. Sensitization about procedures, agencies involved, institutional help, single window clearance.

2. Skill development including specialized short term training
3. Assured supply of raw materials
4. Easy availability of credit on soft terms
5. Special incentives including subsidy in purchase of inputs, tax relief etc.
6. Market support including market intelligence
7. Easy access to suitable technology
8. Exposure to latest technologies through visits/study tours/consultancy
9. Awareness about standard Hygienic Food Manufacturing Practice (GMP) and Hazard Analysis and Critical Control Points (HACCP)
10. Positive approach by the government including commitment to promote food processing.
11. Establishment of a corpus fund for development of agro-processing by women.

Agro-processing Centre (APC)

The model concept of establishing agro-processing centres in production catchments has been evolved in the All India Co-ordinated Research Projects on Postharvest Technology. Agro-processing centre is an enterprise where the required facilitates for primary and secondary processing, storage, handling and drying of cereals, pulses, oilseeds, prices, fruits and vegetables are made available on rental basis to the rural people. Value-added agro-based products and processed food items are also prepared and marketed by the centre. An individual/cooperative/community organization/voluntary organization may manage this type of centre. The centre meets the processing, preservation, handling and marketing needs of the excess produce available in a village or a cluster of villages. Thus, the centre is a means of providing income and employment to rural people through agro-based processing activities of the various products.

Generalized agro-industrial models cannot be suggested for a country as these are location specific. The agro-industrial model for the area could be developed in a manner, which provides sufficient job opportunities to an entrepreneur-unit's adequate profit margin. The APCs also provide good opportunity for women. The research and development institutions as well as commercial houses have developed the agro-processing machines suiting the various requirements.

For developing an agro-industrial model for a particular area, following points should be given due consideration:

1. Existing potential demand of processing produce
2. Availability of raw material *i.e.*, produce
3. Technology/process to be used for processing
4. Volume of production (based on assessment of demand through market survey)
5. Identification of suitable technologies, plants and machinery for the desired volume of production
6. Training facility for operation, repair and maintenance
7. Credit availability at soft rates/subsidies given by the government
8. Facility for storage and marketing.

In order to derive the maximum benefit, the APCs should be properly laid out to optimize available space and utilize the energy efficiently. The selection of equipment (capacity, power requirement etc.) is of utmost importance. Since most of the postharvest operations are seasonal in nature, 2-3 technologies/products may be taken-up to develop an agro-industrial complex to provide regular employment throughout the year. It is estimated that an investment of rupees 300 thousand will create jobs for two persons and generate decent income. With an investment of rupees 500 thousand, it will create employment opportunities for 4 persons and generate income around 90 to 100 thousand rupees.

The Epilogue

It augers well that now some attention is being paid to postharvest management, processing and value-addition of food produce, which hitherto was only marginally mentioned. This has happened when in place of an earlier question asked from the farmers "Why are you not producing?" has given way to "Why are you producing so much?" since more than 60 million tonnes of food grain is stored in godowns and silos. For feeding more than 1 billion population, the country needs to increase the production, no doubt. But when the agriculture on the basis of production alone cannot be sustained as it is becoming more and more un-remunerative, what will be the incentive to the farmers/growers? Many models have been discussed and suggested to solve the issues of competitiveness, lack of income and employment opportunities in the rural areas, which are witnessing large-scale migration to urban areas creating other sets of problems.

We may also have to battle out the problem of huge losses, which makes many holes in the individual and national pockets. Apart from quantitative losses, there are qualitative losses, which reduce the value of raw produce and seriously affect the quality of processed products as well. In this scenario, farmers or consumers and of course the country is big losers. We must set our house in order lest the competitors from outside exploit our weaknesses and make money at our cost. Less income to farmer naturally means reducing his capacity to put more money in the form of inputs and technology which results in lesser productivity–a vicious circle that needs to be busted.

The processing and value-addition of farm produce hold a promise which needs to be fully exploited through development and transfer of cost-effective, user-friendly, energy-saving technologies which are also eco-friendly. A large number of such technologies have been developed in the National Agricultural Research System (NARS) encompassing institutes of the Indian Council of Agricultural Research, State Agricultural Universities and the private sector. The rice, pulses and oil milling industries which are poor in technology are highly polluting, inefficient and result in poor milling out turn. Efforts need to be initiated from pre-harvest level by evolving processable varieties to safe-harvesting without any mechanical damage to safe handling, transport and safe-storage and later processing for making top quality products at par with the best anywhere in the world. A close watch has to be kept on the quality of raw material at all levels till processing is followed by adequate and safe packaging. The codes and standards are becoming very important under the World Trade Organisation Agreement and all concerned (from producer to processor to consumer) need to be sensitized about the GDP, GMP and HACCP.

The agro-processing centre model has presented a tremendous opportunity for ushering in a silent revolution in rural areas especially for poor rural women who are being encouraged to start their own micro-enterprise. The processing and value-addition have the potential to earn the country the agro-dollars, many times more than the petro-dollars. Let this opportunity be fully comprehended and grabbed.

References

Ilyas, S.M. and Goyal, R.K. 2002. Quality assurance in food processing–impact of WTO, In Proc. National Seminar on Development of Food Processing Industries in U.P., held at Lucknow during January 16-17.

Ilyas, S. M., Wanjari, O. D. and Goyal, R. K. 2002. Prospects of income and employment generation through agro-processing and value-addition. Presented at Conference on Emerging Opportunities of Agro and Food Processing Industries in Jharkhand held at Indian Institute of Coal Management, Ranchi during October, 8-9.

Ilyas, S. M. 2003. Importance of postharvest processing and value-addition for additional income and employment generation. Presented at National Symposium on Resource Management for Eco-friendly Crop Production, held at CSAUA and T, Kanpur, February, 26-28.

Kurien, P. P. 1977. Grain legume milling technology. International Proceedings of FAO Expert Consultation on Grain Legume Proceeding. Rome, Italy, AGS/GLP/77/11.

Narsimha, H. V., Ramakrishnan, N., Pratape, V. M., Sashikala, V. B. anf Narsimhan, K. S. 2000. Status of pulse milling industries in India. Presented in Workshop on Recent Development in Pulse Milling, CFTRI, Mysore on 14 February.

Singh, G. 1997. Data book on Mechanization and Agro-Processing Since Independence, CIAE, Bhopal–462038.

Chapter 3

Potato Processing

R. Ezekiel

Head, Division of Crop Physiology and Post harvest Technology,
Central Potato Research Institute, Shimla – 171 001 (HP)

In India potatoes have been utilized largely for consumption as fresh potatoes. Small quantities of potatoes have been processed in the un-organized sector, mostly at the village level. During the last two decades, potato processing in organized sector was taken up in a big way and the increasing popularity of processed products of potatoes like potato chips has led to considerable growth in potato processing industry. Potato production in the country has increased tremendously and this has resulted in frequent glut situations. Potato processing has become necessary in order to utilize the surplus potatoes available and to avoid wastage.

Potatoes can be processed into a variety of products. Potato strips and dice can be produced by simple methods. Potato flakes and granules can also be prepared. Potato flour is another important product from potato, which can be used in baking industry, baby foods and as thickener and flavouring agent in soups and sauces. Potatoes can be processed into French fries, which is the most popular processed potato product in the world. Next to French fries, potato chips are consumed in large quantities throughout the world. While 60 per cent of the total quantity of potatoes produced in U.S.A. is processed, it is only less than 1 per cent in India. Potato chips is the most common and popular processed product of potato in the country.

Quality Requirements for Processing

Chip colour, crispiness and taste determine the quality of potato chips. Light or light golden yellow colour is preferred while brown or black chips are considered undesirable. For chips to be acceptable, besides being light in colour, they should be crispy. Not all potatoes are suitable for chipping. For producing good quality chips, potatoes should meet certain quality requirements. The potatoes should possess required morphological characters. They should be round to round-oval in shape with fleet eyes. The size preferred is 40-60 mm.

Two important characters that determine the processing quality of potatoes are dry matter and reducing-sugar content. Dry matter content of potatoes is important because it determines the yield of the processed products and the quantity of oil consumed while frying. Higher the dry matter content, greater will be the yield of the product and lesser will be the oil consumption. Potatoes with dry matter content of around 20 per cent or more are preferred for the preparation of chips, French fries and dehydrated products.

Reducing sugars (glucose and fructose) are important because they determine the colour of the fried product. Potato chips have light colour if the reducing sugar content is less than 150 mg/100g fresh weight of tuber. If the reducing sugar content exceeds this limit the fried products become dark brown and are unacceptable. This discolouration is due to the reaction called Maillard reaction, between reducing sugars and amino acids, which takes place during the high temperatures of frying. Colour cards have been developed by Central Potato Research Institute (CPRI), Shimla to score chip colour after frying and this gives an idea about the quality of chips that can be obtained from a given lot of potatoes. These colour cards have scores of 1-10 and higher the colour score, more is the browning (Figure 3.1). A colour score up to 4 is acceptable while 5 and above is unacceptable.

Indian Potato Varieties Suitable for Processing

All the Indian potato varieties released until 1998, were developed for table purpose. However, some of these varieties were

Figure 3.1: Colour Cards for Potato Chips

also found suitable for processing since these varieties possess the required morphological characters, high dry matter content and an acceptable level of reducing sugars. All the potato varieties released by CPRI and being grown in the country were tested for the processing quality and three of them were found to be suitable for processing. They are *Kufri Chandramukhi, Kufri Jyoti* and *Kufri Lauvkar*. The potato processing industry in India is mostly using these three varieties. In addition to these three, the institute has developed two new varieties exclusively for processing *viz. Kufri*

Chipsona-1 and *Kufri Chipsona-2*. With these potato varieties being grown in the country on large areas, availability of suitable raw material for processing is no longer a constraint for increasing potato processing in the country. Tuber characters and processing quality of these varieties are shown in Table 3.1.

Table 3.1: Tuber Characters and Processing Quality of Some Indian Potato Varieties

Variety	*Shape/Size*	*Dry Matter (Per cent)*	*Reducing Sugars (mg/100g fresh wt)*
Kufri Chipsona-1	Oval/Large	21-24	45-100
Kufri Chipsona-2	Round/Large	21-25	44-93
Kufri Jyoti	Oval/Large	18-21	106-275
Kufri Lauvkar	Round/Large	18-20	200-250
Kufri Chandramukhi	Oval/large	18-20	250-324

Areas for Growing Potatoes Suitable for Processing

On a large scale trial conducted over several years, the dry matter content of potatoes of several varieties grown at different locations in the country was determined and it was found to vary with the variety and location. Generally, potatoes grown in Punjab had lower dry matter content (less than 20 per cent), while potatoes grown in Bihar and Maharashtra had higher dry matter content (more than 20 per cent). It was found that potatoes grown in warmer areas *i.e.* southern and eastern parts of the country had relatively higher dry matter and lower reducing sugar content. Based on the temperatures prevailing in different parts of the country during the crop growth and their influence on the sugar content, areas suitable for growing processing potatoes have been identified. Generally, the dry matter content was higher and the reducing sugar content was lower in the eastern and the southern regions.

Potato Chips

Potato chips were first prepared in the United States of America during the middle of the nineteenth century. They were originally known as "*Saratoga chips*" and were prepared by housewives. Large scale manufacture of chips began during the Second World War and since then the production and popularity of potato chips has increased tremendously. Today potato chips are popular throughout the world and in terms of the quantity consumed, it is second only to "French fries", which is the most popular processed product of potatoes. The name "chips" was first used by the Americans whereas the British called them "crisps". The American name has been accepted by most of the countries in the world and the word chips is used all over the world, with the exception of the United Kingdom, to refer to thin potato slices fried in oil. Potato chips as manufactured in modern potato processing units is a relatively recent introduction into India, though they have been popular for a long time. Potato chips have been prepared by the housewives in India by blanching potato slices in boiling water for 1-5 minutes, and drying them in the sun. The sun-dried potato chips used to be stored and fried in small quantities, just before consumption. The quantity of potato chips produced by this sector is not comparable to those being produced by the un-organized sector.

Method for Making Potato Chips

Potato chips can be prepared following the procedure given below (Dinesh Kumar *et al.*, 2002).

Potatoes → Washing → Peeling and Trimming → Slicing → Washing → Blanching → Cooling → Drying → Frying → Packing

Peeling

After harvesting or taking out of store, potatoes are washed thoroughly with clean water and then peeled. Normally abrasive peelers are used for peeling. Abrasive peelers use abrasive grits to grind the skin from the potatoes as water sprays flush away the loosened practicles of peel. Peeling losses depend upon the morphological characters of the potatoes. In shallow eyed tubers, peeling losses are much lesser than deep eyed tubers. On an average peeling losses vary from 2 to 5 per cent in Indian varieties. Peeling losses are minimum with smooth potatoes. In small scale processing units, although the peelers may be power operated but the operations of charging and discharging of the peeler is done manually. Batch type abrasive peelers are used in this case. In big processing units, feeding of tubers to peelers, peeling along with washing and then discharging of the peeled potatoes are all done automatically.

Trimming

Peeled and washed potatoes require hand trimming. This is done to remove residual skin, eyes and green portions. The amount of trimming depends upon the product requirements.

Slicing

Peeling is followed by slicing to a desirable thickness. Potato slices should have uniform thickness and smooth surface with minimum cell fracture. Slices with rough surfaces absorb higher amount of oil while frying. It is necessary to wash the slices in water, to remove the adhering starch and other particles. In large chip processing plants, slicing and washing processes are done simultaneously. Starch removal from surface ensures that chips will not stick to each other while frying. Washing also helps in the removal of some sugars and this helps in improving the quality of chips in respect of colour and taste. CPRI has developed a slicing machine. It is a manually operated machine, which is used for producing potato slices of desired thickness. This machine can be operated by one person and the output is 50 kg/h.

Blanching

Blanching is done to improve the chip colour when sugar level in potatoes is very high. In this process slices are given a dip in hot water (60-80°C) for 2-3 minutes. Blanching removes some quantity of soluble constituent, including reducing sugars from potatoes. Blanching also inactivates some of the enzymes which cause enzymic discolouration. After blanching, slices are partially dried to remove excess moisture. This can be done with paper towel or centrifugal driers or by using heated air. Removal of surface moisture helps in shortening the frying time. After the drying process, slices are ready for frying.

Frying

Slices are fried immediately after drying. If delayed, slices may turn brown thereby affecting chip colour and crispiness. Batch frying method is commonly used. Frying of chips is done at 180-190°C till bubbling stops. For small scale processing, electrically or gas heated friers are used in which temperature is indicated continuously. The oil content of chips is relatively high in this method of frying because of decrease in oil temperature during the frying process, when slices are discharged in the oil. In modern processing plants continuous friers are used in which slices are introduced at 185°C from one side and leave the frier at 176°C. Some of the advantages of continuous friers are that the oil does not get overheated, the fuel consumption is less and the quality of chips is better. For domestic uses, small electric operated friers of 2-3 litre capacity in which temperature control is provided can be used. Frying can also be carried out in conventional "Kadahi" but chips of uniform quality cannot be produced since there is no temperature control.

Any type of good quality oil can be used for frying keeping in mind the cost of frying medium. Hydrogenated "ghee" should be avoided in winter as it affects taste and texture of chips. For one kg of chips about 275-350g oil is consumed depending upon the dry matter of potatoes and frying temperature. Oil should be changed after 6-8 hours of continuous use. Fried chips are taken out of the frier and kept for a few minutes in a sieve to drain the excess oil. In industry, automatic oil extracting systems are used for reducing oil content of chips.

Effect of Storage Conditions on Processing Quality

Potatoes are normally stored in cold store maintained at 2-4°C. This storage temperature is ideal for seed potatoes because sprouting does not occur at this temperature. But for storage of potatoes meant for processing, this is not the ideal storage temperature because at such low temperature (<5°C) a part of starch stored in tubers is converted to reducing sugars that makes the potatoes unsuitable for processing. By a process called reconditioning, which involves storage of potatoes at high temperatures, the level of reducing sugars, could be brought down considerably. However, reconditioning does not help in achieving the original level of reducing sugars, at the start of the storage period. Potatoes meant for processing purposes are stored at 8-12°C in European and American countries. However, when stored at this temperature, potatoes sprout. Therefore, it is necessary to treat the potatoes with a sprout suppressant. The commonly used sprout suppressant is CIPC (Isopropyl N-chlorophenyl carbamate). It is necessary to check sprout growth during storage because sprouting increases storage losses and results in shriveled tubers.

Chipping Quality Characteristics of Different Varieties Stored in Different Storage Systems

Two varieties *viz. Kufri Jyoti* and *Kufri Bahar* stored at room temperature, in heaps at 10°C, and 2-4°C were analysed for chipping quality after 90 days of storage at Modipuram. The experiment was started in the first week of March 2001 and terminated in the first week of June 2001. Chip colour was acceptable at room temperature and in heap storage, while in cold stored and 10°C stored potatoes, chips were dark brown (colour score up to 10 was found). Dry matter content increased marginally in all the storage systems. Reducing sugar content was the highest in cold stored potatoes in both the varieties as shown in Table 3.2.

Table 3.2: Chipping Quality of *Kufri Jyoti* and *Kufri Bahar* After 90 Days of Storage at Room Temperature, in Heaps, at 10°C and at 2-4°C.

Variety	*Parameter*	*At Zero Day*	*After 90 Days at Room Temp.*	*After 90 Days in Heap*	*After 90 Days at 2-4°C*	*After 90 Days at 10°C*
Kufri Jyoti	Chip colour	6.0	4.0	3.6	10.0	6.0
	Dry matter (per cent)	18.2	21.1	21.2	18.0	17.9
	Reducing sugars (mg/100g)	245	18	43	675	263
	Sucrose (mg/100g)	117	47	344	229	107
Kufri Bahar	Chip colour	8.0	5.0	6.5	10.0	10.0
	Dry matter (per cent)	18.7	22.4	24.3	21.6	22.4
	Reducing sugars (mg/100g)	254	88	139	810	226
	Sucrose (mg/100g)	251	366	767	467	175

Dehydrated Potato Products

Processed potato products like potato chips are no doubt popular but the cost of production of potato chips is very high because of the sophisticated technology used and high cost of frying medium and electricity. Therefore these processed products are not within the reach of common man. In order to make the processed potato products available to a large section of the population at an affordable price, we need to develop simpler methods of potato processing.

Preservation of foods by drying is perhaps the oldest method known to man. Drying results in the lowering of moisture content leading to lesser chances of microbial growth. As a result, the product has a longer shelf-life. The reduction in moisture content is accompanied by a reduced bulk which facilitates storage, transportation and packaging. Potato slices can be dehydrated in the sun, stored, fried and consumed when required. Dehydrated potato chips is the most common processed potato product in the rural and semi-urban areas in the country. Solar dehydration of potatoes is very popular in several states like Maharashtra, Gujarat, U.P. and M.P. and remains a viable alternative to expensive and sophisticated potato processing technology.

Dried Potato Slices

Dried potato slices are also known as raw chips, which are consumed after frying and salting. These are prepared at home by housewives and also by small scale manufacturers. The housewives simply hand peel the potatoes, cut them into slices and then boil the slices in water for 5-8 minutes. The slices are dried in the sun till they become brittle. Small scale food processors generally use electrically operated slicers to make 1-2.5 mm thick slices of potatoes. These slices are washed in cold water to remove adhering starch. The slices are then blanched in hot water (80-90°C) containing potassium metabisulphite. The blanched slices are then dried in the open. Discolouration of potato slices is minimized by immersing them in cold water and blanching inactivates the enzymes responsible for browning and thus reduces discolouration. It is recommended that 2 mm thick potato slices can be dried quickly by spreading on a black polythene sheet and varieties like *Kufri Sindhuri* having higher dry matter content can be used to get high yield of dehydrated products.

Rate of Dehydration

The potato slices dried faster when both the sides were exposed to the sun alternately as compared to only one side (Figure 3.2A). However, the differences narrowed down as the drying progressed

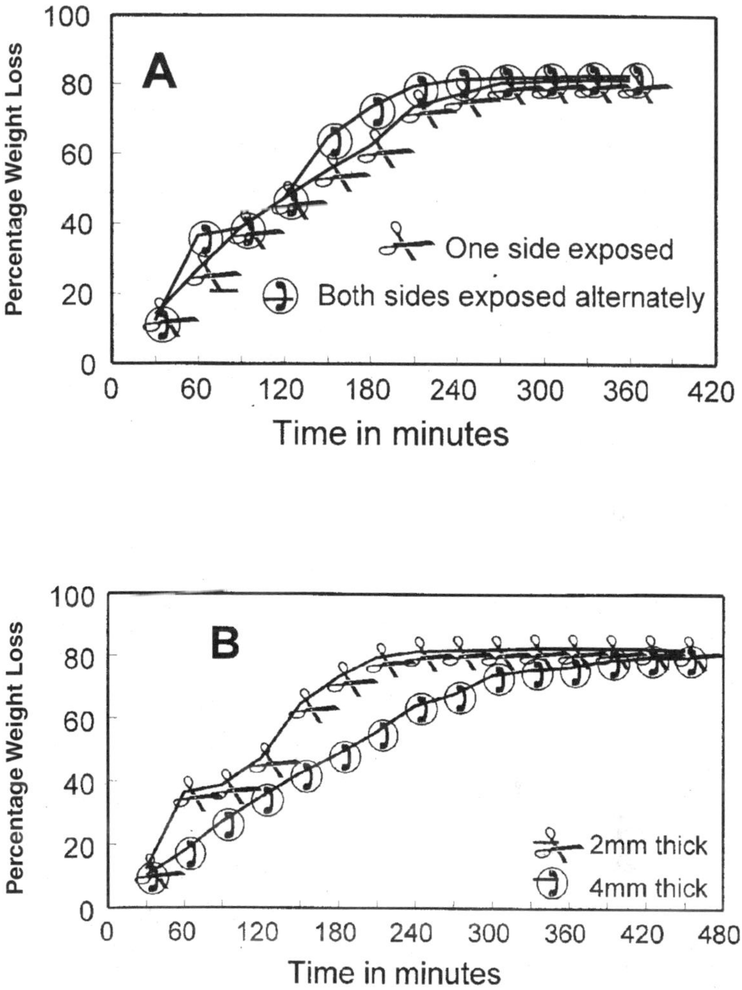

Figure 3.2: Weight Loss in Potato Slices of K. Jyoti (A) Different Sides of 2mm Thick Slices Exposed; (B) Slices of Two Thickness

and there was little difference after 300 minutes when complete drying was achieved in both the cases. In 2mm thick slices, 80 per cent weight loss occurred in 240 minutes of drying whereas in 4 mm thick slices it was about 60 per cent (Figure 3.2B). Rehydration was better in boiling water as compared to soaking in water at room temperature. In boiling water, the rate of rehydration was up to 450 per cent whereas, at room temperature it was only 188 per cent (Figure 3.3A). Weight gain was better in 2mm thick dehydrated slices. In 2 mm thick slices, the rate of rehydration was 485 per cent after 105 min. of soaking whereas in 4 mm thick slices it was 400 per cent (Figure 3.3B).

The rate of dehydration was rapid in the first one hour and it declined thereafter. Though exposing both the surfaces alternately helps in faster drying, considering the additional labour involved in turning the slices frequently, it does not offer much advantage. The higher rate of rehydration of up to 450 per cent in 2 mm thick dehydrated slices shows that it is possible to reconstitute solar dehydrated slices by soaking them in boiling water for 45 minutes. The reconstituted slices can be mashed and used in preparations like stuffed parathas. The cubes take a long time to dry in the open sun and rehydration of cubes was not very satisfactory. It may be concluded that solar dried potato chips can be prepared, conveniently stored and utilized in two ways. They can be fried when needed and consumed as chips or they can be reconstituted, mashed and used during summer months when the price of potatoes is quite high.

Preparation of Dehydrated Potato Chips Using Solar Dryer

CPRI has also developed two types of solar dehydrators for sun-drying potato slices. One is a large cabinet with a solar concentrator and air temperature inside is 13-21°C higher than ambient air. The other is a small, domestic solar dehydrator. These two solar dehydrators can be used to dehydrate potato slices in 4-6 hours.

Dehydrated chips from three different varieties were prepared using solar dryer and open sun drying at Modipuram. The values presented are the average of 3 separate determinations of 2 kg potatoes each. There was very little difference between the two methods of drying in the quality of dehydrated chips (Table 3.3).

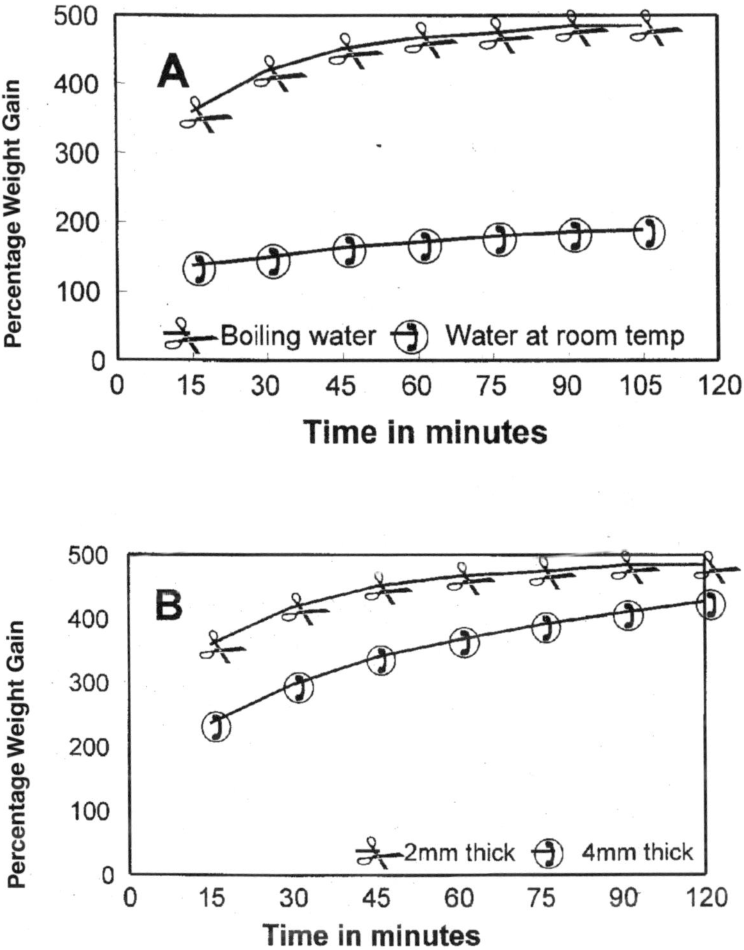

Figure 3.3: Weight Gain in Dehydrated Potato Slices of K. Jyoti (A) Soaked In Cold and Boiling Water; (B) Slices of Two Thickness

Table 3.3: Chipping Quality Characteristics of Dehydrated Chips Prepared from *Kufri Lauvkar* Using Solar Dryer (SD) and Open Sun Drying (OSD)

Type of Drying	*Yield (Per cent)*	*Chip Colour*	*Oil Uptake (Per cent)*	*Texture*	*Taste*
Solar drying	14.4	3	15.8	Crunchy	Very good
Open sun drying	15.1	3	16.2	Crunchy	Very good

Nutritional Losses During Processing

Nutritional losses in dehydrated chips prepared from *Kufri Jyoti* varied from 73 to 86 per cent (Table 3.4). Maximum loss of 86 per cent was observed in soluble protein and the loss in reducing sugar was 81 per cent. Minimum loss of 73 per cent was observed in free amino acids. In cubes, the losses in reducing and total sugars were 73 and 78 per cent respectively in *Kufri Jyoti*, and in *Kufri Chandramukhi*, the losses were 77 and 75 per cent respectively (Table 3.4). The losses in soluble protein was quite high at 77 per cent and the losses in free amino acids and total phenol were also above 70 per cent. During preparation of several dehydrated products from potatoes, losses of 18 per cent in glucose, 90 per cent in fructose, 64 per cent in sucrose, 26 per cent in protein and 48 per cent in amino acids have been reported (Lisinska, 1989).

Table 3.4: Nutritional Losses During Preparation of Solar Dehydrated Potato Chips

Parameter	*Variety*					
	Kufri Jyoti			*Kufri Chandramukhi*		
	Raw Tuber	*Dehydrated Chips (g/100g dry wt.)*	*Per cent Loss*	*Raw tuber*	*Dehydrated Chips (g/100g dry wt.)*	*Per cent Loss*
Total sugars	2.50	0.54	78	1.97	0.45	77
Reducing sugar	1.70	0.32	81	1.25	0.31	75
Soluble protein	1.97	0.27	86	1.15	0.26	77
Free amino acids	0.34	0.09	73	0.14	0.04	71
Total phenol	0.38	0.09	76	0.24	0.06	75

French Fries

French fries are very popular in the developed countries and in U.S.A. French fries form 43 per cent of the processed products from potatoes (Talburt and Smith, 1987). In India the quantity of French fries consumed is much less compared to potato chips. However, the demand for French fries is increasing gradually. Generally, frozen French fries (par-fries) are prepared and sold. These are finish fried in deep fat. Potatoes are peeled, trimmed and cut into sticks. The cross section can range from 7 to 12 mm and is generally 10 mm. The length of sticks again varies and is normally 5 to 7 cm. Slivers (thin slices) and nubbins (short or broken pieces) are removed as wastage. The French fry sticks are blanched, par- fried and frozen. Finish frying is done for about 2 minutes at 180°C and served hot. French fries can also be prepared from fresh potatoes at home and restaurants by frying them in deep fat in two stages.

Preparation of French fries

French fries were prepared from tubers of *Kufri chipsona-1*, *Kufri Chandramukhi*, *Kufri Jyoti*, *Kufri Bahar* and *Kufri Anand*. Peeling losses were maximum in *Kufri Anand* but wastage was minimum (Table 3.5). The yield of French fries was high in *Kufri Anand* followed by *Kufri Chipsona-1* and *Kufri Jyoti*.

Table 3.5: Processing Losses During the Preparation of French Fries

Variety	*Peeling Loss (Per cent)*	*Wastage (Per cent)*	*Yield of Sticks (Per cent)*	*Yield of French Fries (Per cent)*
Chipsona-1	7.5	12	79.5	43.5
K Chandramukhi	8.0	17	74.3	39.7
K. Jyoti	7.6	15.8	75.5	41.8
K. Bahar	8.6	16.7	74.6	39.6
K. Anand	8.8	9.8	80.9	44.1

Freshly cut sticks were fried for 10 minutes. While freshly prepared fries were crisp and highly acceptable in the first four varieties, they were unacceptable in *Kufri Anand* (Table 3.6). After 5 minutes of frying the fries became soft in all the varieties. The colour of French fries produced from *Kufri Chipsona-1* was highly acceptable with soft texture, good taste and minimum browning. French fries produced from *Kufri Chandramukhi* (KCM) and *Kufri Jyoti* were also highly acceptable, while French fries produced from *Kufri Bahar* were just acceptable. French fries prepared from *Kufri Anand* were unacceptable.

Table 3.6: Quality of French Fries Prepared from Five Potato Varieties

Variety	*Colour*	*Texture*	*Taste*	*Flavour*	*Browning (Per cent)*	*Overall Acceptability*
Chipsona-1	HA	Soft	Good	Normal	15	HA
KCM	A	Soft	Satisfactory	Normal	25	HA
K. Jyoti	A	Soft	Satisfactory	Normal	18	HA
K. Bahar	A	Soft	Satisfactory	Normal	25	A
K. Anand	UA	V. Soft	Satisfactory	Typical	100	UA

A: Acceptable; HA: Highly acceptable; UA: Unacceptable.

After frying *Chipsona-1,Kufri Chandramukhi, Kufri Jyoti* and *Kufri Bahar* was highly acceptable and crisp except K. Anand. After 5 minutes all the varieties got soft. Fresh fries were fried for 10 minutes. Based on the above results *Kufri Chandramukhi, Kufri Jyoti* and *Kufri Chipsona-1* are recommended for French fries.

Oil Content of French Fries

French fries were prepared from three potato varieties. The sticks were par fried and stored frozen for 90 days. There was little difference in the oil content of French fries after 30, 60 and 90 days of storage (Table 3.7).

Table 3.7: Oil Content of Frozen French Fries Prepared from Three Varieties

Variety	*Days of storage*		
	30	*60*	*90*
Kufri Jyoti	14.0	11.6	13.8
Kufri Chipsona-1	18.2	18.9	14.2
Kufri Chipsona-2	14.6	14.5	15.2
Mean	15.6	15.0	14.4

Potato Starch

Potato is rich in starch with 14-16 per cent starch content. Although potato starch is costlier when compared to maize and cassava, because of its superior properties it is preferred for special applications in pharmaceutical, textile, food and paper industries. The demand for potato starch is because of its unique properties (Talburt and Smith, 1987). The advantage of using potatoes for starch production is that substandard potatoes not suitable for table or processing can be used for starch production. Potatoes, which are too small, too large and damaged, can also be used in starch manufacture.

Starch Extraction Procedure

The procedure for starch extraction is described below.

Potatoes

↓

Wash and peel

↓

Cut into small pieces

↓

Dip in water containing 1 per cent potassium metabisulphite

↓

Grind in a grinder

↓

Sieve it through muslin cloth

↓

Allow the starch to settle

↓

Decant the starch 2-3 times with water till the water is clear

↓

Dry it in oven at 60°C or in the sun
(If dried in the sun protect from dust)

↓

Grind and sieve it to get fine powder

Washing and Peeling

Tubers should be washed to remove all the adhering soil particles followed by trimming/peeling to remove damaged portions of potato. Potatoes are peeled with the help of peelers and they are kept in water after peeling.

Disintegration

Starch granules can be liberated from potato tubers by disruption of the cell walls. This is done by potato disintegration, *i.e.*, rasping. Peeled potatoes are sliced into slices of thickness 1-2 mm and keeping them in water. These slices are crushed in excess water (5 times) in a grinder (4-5 min.) to make the slurry.

Second Extraction

The residue is suspended in excess water containing 1 per cent potassium metabisulphite. It is crushed again in a grinder for 4-5 minutes. The second extraction is done to improve starch recovery. It is sieved through muslin cloth and the filtrate is allowed to settle in the tube along with the filterate obtained after first extraction.

Sieving

Starch is contaminated with fine fibrous substances, mineral compounds, sugar and nitrogenous compounds. All impurities should be removed. This is done by washing starch with excess of water (several times). Each time supernatant is discarded and settled starch resuspended. This is repeated 2-3 times to get the clean starch.

Drying and Packaging

Starch is dried in an oven at 40°C. Before complete drying, it is made into a fine powder and then dried completely. Drying can also be done in the sun. This dried starch can be packed in polythene bags. Starch has long shelf-life.

Project Cost and Plant Machinery

A small-scale plant will require equipment costing Rs. 3.5-4 lakh excluding land and labour cost. The equipment required includes Washer, Peeler, Slicer, Pulverizer, Settingtanks (Steel), Bucket centrifuge, Dryer, Miscellaneous (racks balance, packing machine etc.)

All these equipment are available indigenously. The cost of potato starch production, after taking cost of raw material, equipment and processing into consideration works out to Rs. 40-50/kg. If the damaged, cut, cull and rejected tubers are used as the raw material, the cost of starch production can be further reduced.

Suitability of Indian Potato Varieties for Starch Production

Potatoes with high dry-matter content are preferred for starch production since high dry-matter content leads to higher starch yield. On an average, Indian potato varieties contain about 20 per cent dry-matter and about 14 per cent starch. When grown under warmer temperature of penisular and eastern India, the dry-matter content goes up to 22 per cent with a corresponding increase in starch content. In countries like the Netherlands, starch varieties have been developed exclusively for the production of potato starch. These varieties have a starch content of 18 per cent or more. Potatoes with starch content of 13 per cent are also used for starch production in USA. Indian varieties with an average starch content of 14 per cent are quite suitable for producing starch. Cultivars which have been found suitable for starch extraction are *Kufri Sindhuri, Kufri Lauvkar, Kufri Lalima, Kufri Chandramukhi, Kufri Badshah* and *Kufri Jawahar*. The two new potato varieties, *viz. Kufri Chipsona-1* and *Kufri Chipsona-2*, developed exclusively for processing purposes contain 22 per cent dry matter when grown in North Indian plains. These two varieties can be used to get higher starch yield.

Studies conducted at CPRI, Shimla have shown that most of Indian potato varieties have an average granule size of 40-60 µm (Table 3.8) and these potato varieties do not show difference in starch granule distribution. Amylose content ranges between 19.3 and 21.7 per cent. Starch yield was found to be directly correlated to dry-matter content of the tubers. Phosphorus content in potato starch was found to vary between 0.06 and 0.08 per cent which was much higher compared to other tuber crop starches.

Potato Processing in the Organized Sector

Potato chips production by the organized sector increased rapidly after the introduction of the liberalization policy of the government of India. The organized sector in the country produces about 30,000 tonnes of potato chips per year. Popular brands like Uncle chips and Haldiram are located in Delhi or around Delhi (Ghaziabad). Delhi alone has four major potato processors *viz.* Haldiram, Kaleva, Wah Bakana and Twinkle with a total production of more than 20t of chips per day. Uncle Chips unit is located in NOIDA, (Ghaziabad). The branded potato chips have approximately 70 per cent share of the market in Delhi while the potato chips produced by the unorganized sector account for about 30 per cent of the market (Kamal Jame *et al.*, 2001). The second popular product is potato *lachha* and this market is also dominated by branded manufacturers. *Aloo bhujia*, which contains potato only as one of the ingredients, is also quite popular and this market is dominated by Lehar and Haldiram. The potato processors in the organized sector normally procure potatoes from Indore, Hassan, Belgaum and Nagrota. The major constraints faced by the potato processors are non-availability of quality raw material from November to January and transporting potatoes suitable for processing over long distances (1000 to 2000 km) at considerable cost.

Table 3.8: Properties of Starch Extracted from Indian Potato Cultivars

Cultivars	*Dry Matter (%)*	*Starch Content (%)*	*Granule Size (μm)*	*Amylose Content (%)*	*Moisture Content (%)*	*Phosphorus Content (%)*	*Viscosity (cP)*
K. Chandramukhi	22.1	16.3	59.6	20.2	16	0.06	448
K. Jyoti	19.2	14.4	52.5	21.1	17	0.07	661
K. Sindhuri	21.7	15.9	45.0	21.7	19	0.07	376
K. Lalima	22.2	16.3	45.0	19.3	16	0.08	350

Potato Processing in the Unorganized Sector

Potato processing in the unorganized sector is of considerable importance in a country like India where majority of the population cannot afford to purchase potato chips produced by the organized sector which costs around Rs. 300 per kg. Potato chips produced by the unorganized sector costs Rs. 150 to Rs. 200 per kg. In Kolkata alone there are about 200 small units producing processed products from potatoes. Small scale potato processors are found in almost all cities and towns in the country and they produce a variety of products like potato chips, potato lachha, potato bhujia, potato pops, potato papads, dehydrated potato chips etc. Dehydrated potato chips and potato flour are being produced by several small scale potato processors in and around Bareilly.

Although the price of the potato chips produced by the unorganized sector is low, the quality of the chips is poor as compared to that produced by the organized sector (Manju Rani and Ezekiel, 2001). The chips produced by the unorganized sector can have up to 80 per cent browning, 50 per cent broken pieces and 10 per cent green chips.

References

Dinesh Kumar, Manish Das, Kamal Jame and Ezekiel R. 2002. Processing of potatoes. *Potato, CPRI, Newsletter*. 23, 23-27.

Kamal Jame, Dinesh Kumar, and Ezekiel R. 2001. Market status of different processed potato products in Meerut, Ghaziabad and Delhi. *Journal of Indian Potato Association*. 28, 169-171.

Lisinska, G. 1989. Dehydrated potato products. In G. Lisinska and W. Leszczynski, (Eds), Potato Science and Technology (pp. 234-80). *London: Elsevier Applied Science.*

Manju Rani and Ezekiel, R. 2001. Potato processing in unorganized sector and quality of potato chips available in the market. *Journal of Indian Potato Association*. 28, 159-161.

Talburt W. F and Smith, Ora. 1987. Potato Processing. New York: Van Nostrand Reinhold.

Chapter 4

Postharvest Management

M.K. Garg

Department of Agricultural Processing & Energy,
CCS Haryana Agricultural University, Hisar – 125 004

Postharvest management is crucial for self sufficiency and development of the country. A substantial percentage of food grains, fruits, vegetables and dairy products are lost every year due to poor postharvest management in India which results in the loss of more than Rs. 50,000 crore per annum. This article describes different postharvest unit operations for grains, fruits and vegetables which can contribute in the reduction of losses besides ensuring food safety and quality.

Every day, millions of individuals do not have enough to eat, and many of them simply do not eat at all. In the developing countries, food self-sufficiency still remains an unachieved objective. This is not always or only because of the inefficiency of local production systems, but, the extent of postharvest losses sometimes seriously limits the impact of efforts made to increase food production. To feed ever increasing population, there is a need of proper postharvest management.

India produces around 78 million metric tonne of rice against the world's production of 540 million metric tonne thus accounting for around 14 per cent of rice production in the world. Wheat accounts for around 30 per cent of the total cereal production in the world.

More than 6 per cent of rice is lost due to poor storage design/ practices. The bulk storage facilities existing in India are inadequate to meet the increasing storage requirements.

Total production of fruits in the world is around 370 million metric tonne. India ranks first in the world with an annual output of 32 million metric tonne. India with its current production of around 32 million metric tonne accounts for about 8 per cent of the world's fruit production. The diverse agro-climatic zones of the country make it possible to grow almost all varieties of fruits and vegetables. Although India is the largest producer of fruits in the world, the availability per capita is only about 100 grams per day. However, it is estimated that more than 20-22 per cent of the total production of fruits is lost due to spoilage at various postharvest stages. Thus the per capita availability of fruits is further reduced to around 80 grams per day which is almost half the requirement for a balanced diet.

India is also the second largest producer of vegetables in the world (ranks next to China) and accounts for about 15 per cent of the world's production of vegetables. The current production level is over 71 million metric tonne and the total area under vegetable cultivation is around 6.2 million hectares which is about 3 per cent of the total area under cultivation in the country. It is estimated that around 20-25 per cent of the total vegetable produce lost due to poor postharvest practices. Less than 2 per cent of the total vegetables produced in the country are commercially processed as compared to 70 per cent in Brazil and 65 per cent in USA.

Postharvest management involves all the activities that occur after the production of agricultural commodities including procurement, packaging, storage, transportation, processing and marketing of agricultural products from the farm gate to the distributor and the sequences and interaction of these activities contribute to the formation of a system known as postharvest management. Improved postharvest management depends on the quality and efficiency of handling, processing and preservation techniques used. Its objectives are:

1. To ensure food safety and quality.
2. To act as source of employment.
3. To meet emergent form of consumers' demands.
4. Value-addition to products.

5. To raise living standard of the farmers in particular.
6. To reduce losses after harvesting.

Postharvest Management of Grains

Postharvest operations of grains are shown in Figure 4.1. The management during these operations is important in order to maintain the quality and reduce the losses.

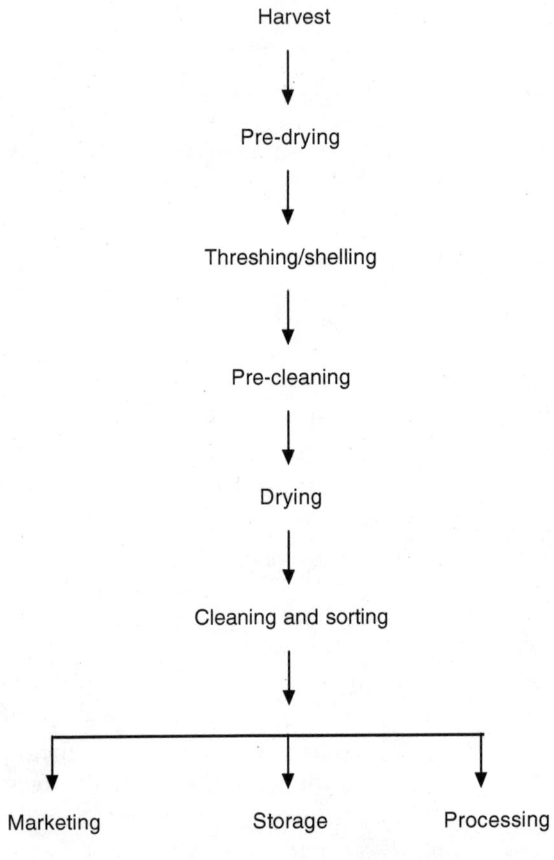

Figure 4.1: Different Unit Operations after Harvesting of Grains

Postharvest system of grains includes following unit operations:

Harvesting

It should not be undertaken until the produce has reached its optimal maturity.

Pre-drying

After the harvest, it may be necessary to pre-dry the produce before the subsequent threshing or shelling operation.

Cleaning

The grain must be cleaned and dried, so that it can be stored or undergone further processing.

Storage

Grain can be stored in bulk or in the bags, at the farm where it was produced, in the collection centres, or with the storage agencies in a scientific manner.

Processing

The grain is sent from warehouses to the markets for sale to the consumers, to small-scale food-processors, or to agro food-industries.

Postharvest losses mean a measurable quantitative and qualitative loss in a given product. These losses can occur during any of the various phases of postharvest system. The direct causes of postharvest losses are premature harvest, poor threshing, insufficient drying, insufficient cleaning, bird attack, rodent attack, insect attack, micro-organism attack, biochemical change, leakage or waste, incorrect moisture content for storage, inadequate storage and transport techniques.

Pre-drying

At harvest time, the cut portions of the plant may contain too much green plant matter while the grain may not have reached a uniform degree of maturity and may have too high moisture content. "Pre-drying" is the stage of the postharvest system during which the harvested product is dried in order to undergo the next operation of threshing or shelling under the best possible conditions.

Cleaning and Grading

It is one of the most important postharvest operations undertaken to remove foreign and undesirable materials and to separate grains into various fractions. Grading of cleaned seeds (cereals and oilseeds) is done to obtain good quality seed for higher germination and viability. A number of cleaners/graders are available commercially.

Drying

Drying of agricultural produce (cereals and oilseeds) is an important unit operation under postharvest phase. It refers to the removal of moisture from grains to a pre-determined 'safe moisture' level. The main purpose of drying is to lower the grain moisture content in order to guarantee conditions favourable for storage or for further processing. It also reduces grain respiration through removal of excess moisture and prevents quality deterioration, and thus reduce losses. After threshing, the moisture content of most grains is too high for good conservation (13-14 percent). Drying makes the material stable and the attack of insects and micro-organisms during storage period can be kept at negligible level. It also removes the volatile materials present inside solid matters.

Storage

Storage means the phase of postharvest system during which the products are kept in such a way as to guarantee food security. This single component in the chain of postharvest handling of grains is responsible for losses to the tune of 7 per cent. Its objectives are:

1. To permit deferred use (on an annual and multi-annual basis) of harvested agricultural products.
2. To ensure availability of seeds for the crop cycles to come.
3. To guarantee regular and continuous supply of raw material for processing industries.
4. To balance the supply and demand of agricultural products thereby stabilizing market prices.

The following rules must be observed for safe storage of grains in the godowns:

1. Avoid filling in more grain than the store house can hold.

2. Store only those products that are very dry and clean.
3. Apply the inventory principle of "First In, First Out" (FIFO).
4. Maintain the cleanliness and good condition of equipment and the stored products.
5. At the time of stacking, provide for individual lots, separated by type of product, quality and date of entry into the store house.

Transport

It generally marks the passage from one stage of postharvest system to the next. In India, significant losses are recorded during transportation of food grains due to improper stitching of bags, multiple handling, and pilferage. In the developed countries, the entire transportation, handling and storage system has been highly mechanized. There the grains are transported in specially designed box cars in bulk quantities and manual handling is minimized. This results in very low spillage losses. Such mechanized devices are not yet practiced in India.

Transport whether traditional or mechanized, is needed to move the agricultural commodities from the field to the drying site; to the farmer's store house or to collection centre warehouses; to processing industries or to bigger central storage buildings and to wholesalers or retailers for final marketing.

> In India, tractors with trailers are generally used. Grains in gunny bags are also transported on trucks of varying capacities.

Postharvest Management of Fruits and Vegetables

Production practices have a tremendous effect on the quality of fruits and vegetables at harvest and on postharvest quality and shelf-life. It is well known that some cultivars have a longer shelf-life than others. In addition, environmental factors such as soil type, temperature, frost, and rainy weather also affect the storage life and quality. Management practices can also affect postharvest quality. Produce that has been stressed by too much or too little water, high rates of nitrogen or mechanical injury (scrapes, bruises, abrasions) is particularly susceptible to postharvest diseases.

Harvest Handling

It is important to harvest fruits, vegetables, and flowers at the proper stage and size and at the peak quality available. Immature or over-mature produce may not last as long in storage as that picked at proper maturity. Harvest should be completed during the coolest time of the day (which is usually in the early morning) and the produce should be kept under shade in the field. Crops destined for storage should be as free as possible from skin breaks, bruises, spots, rots, decay, and any other deterioration. Bruises and other mechanical damages not only affect appearance but also provide entrance to harmful organisms as well.

Postharvest rots are more prevalent in fruits and vegetables that are bruised or otherwise damaged. Mechanical damage also increases moisture loss. The rate of moisture loss may be increased by as much as 400 per cent by a single bad bruise on an apple, and non-skinned potatoes may lose three to four times as much weight as skinned ones. Damage can be prevented by training the work force to handle the crop gently; harvesting at proper maturity; harvesting dry whenever possible; handling each fruit or vegetable no more than necessary (field pack if possible); installing padding inside bulk bins and avoiding over or under-packing of containers.

Packaging

Packaging should be designed to prevent physical damage to produce and be easy to handle. An ideal package of fruits and vegetables should protect fruit from mechanical injury, keep fruits in clean and hygienic manner, serve as an efficient handling unit for consumers and producers, be convenient for transportation, serve as convenient warehouse or storage unit, reduce the wastage, provide adequate ventilation to fruits to avoid storage and transport loss, have height adjustment in such a way that fruits of lower stack do not get pressed by the upper stack, be attractive and have retailer acceptance, fulfill government rules and regulations, if any, and be economical. The outer surface of the package should be easily printable in order to specify name of the product, company, quantity, etc.

Organization for Economic and Co-operative Development (OECD) has recommended external standard base dimensions for individual package as 60 cm × 40 cm; 50 cm × 30 cm or 40 cm × 30 cm. The base dimensions have to be fixed taking into consideration palletisation and dimensions of the pallets. The most effective rationalization can be achieved by using a single type of box of the same dimensions for as large a variety of products as possible, changing only the height of the box if required. Table 4.1 presents the optimum packaging weights/units of different fruits and vegetables when packed in 40 cm × 30 cm box.

Table 4.1: Optimum Packaging Weight/Units for Different Types of Fruits

Fruit	*Net Weight, kg*	*Height of the Box, cm*
Avocado	4	10–12
Baby pineapple	3	10
Fig	2	9
Hot pepper	4	12
Lime	3	8
Mango	4	8
Melon	4	8
Okra	4	10–15
Papaya	2	10–12
Passion fruit		8

Storage Crops

Most of the storage crops require low temperature and high humidity which may not happen easily. Several other crops require low humidity and low temperature. Root crops such as beets, carrots, turnips, rutabagas, and leeks store best at 0°C and 90 per cent humidity. Potatoes prefer temperatures of 5 to 15°C and 90 per cent humidity. Onions and garlic like cool to 0°C but require less humidity (about 65-75 per cent). Winter squash prefers temperatures between 10 to 15°C, but under dry conditions. Whatever the method of storage may be, only perfect produce is suitable for long-term storage. Any damaged produce is going to spoil itself and induce spoilage in the rest of the crop.

Temperature

Temperature is single most important factor in maintaining quality after harvest. Refrigerated storage retards the following elements of deterioration in perishable crops:

1. Aging due to ripening softening and textural/colour changes.
2. Undesirable metabolic changes and respiratory heat production.
3. Moisture loss and wilting.
4. Spoilage due to invasion by bacteria, fungi, and yeasts.
5. Undesirable growth, such as sprouting of potatoes.

One of the most important functions of refrigeration is to control the crop's respiration rate. Respiration generates heat as sugars, fats, and proteins in the cells of the crop are oxidized. The loss of these stored food reserves through respiration means decreased food value, loss of flavour, loss of saleable weight, and more rapid deterioration. The respiration rate of a product strongly determines its transit and postharvest life. Higher the storage temperature, higher the respiration rate will be.

For refrigeration to be effective in postponing deterioration, it is important that the temperature in cold storage rooms be kept as constant as possible. Exposure to alternate cold and warm temperature may result in moisture accumulation on the surface of produce (sweating), which may hasten decay. Storage rooms should be well insulated, adequately refrigerated and should allow for air circulation to prevent temperature variation. Table 4.2 shows the optimum temperature ranges for various crops.

Pre-cooling

Pre-cooling is the first step in good temperature management. The field-heat of a freshly harvested crop *i.e.*, the heat the product holds from the sun and ambient temperature, is usually high and should be removed as quickly as possible before shipping, processing, or storage. Refrigerated trucks are not designed to cool fresh commodities but only maintain the temperature of pre-cooled produce. Likewise, most refrigerated storage rooms have neither the refrigeration capacity nor the air movement needed for rapid cooling. Therefore, pre-cooling is generally a separate operation requiring special equipment and/or rooms. Rapid pre-cooling to the product's lowest safe temperature is most critical for crops with inherently high respiration rates.

Table 4.2: Optimum Storage Conditions for Vegetables and Fruits

Fruit/Vegetable	*Temperature (°C)*	*Relative Humidity, %*	*Pre-cooling Method*	*Storage Life, Days*	*Ethylene Sensitive*
Apples	-1 to 4	90-95	R, F, H	90-240	Y
Apricots	0	90-95	R, H	7-14	Y
Asparagus	0 to 2	95-100	H, I	14-21	Y
Avocados	4 to 13	85-90		14-28	Y
Bananas	13 to 14	90-95		7-28	Y
Beans, snap	4 to 7	95	R, F, H	10-14	Y
Beans, lima	3 to 5	95		7-10	
Beets, root	0	98-100	R	90-150	
Blackberries	-0.5 to 0	90-95	R, F	2-3	
Blueberries	-0.5 to 0	90-95	R, F	10-18	
Broccoli	0	95-100	I, F, H	10-14	Y
Brussel sprouts	0	95-100	H, V, I	21-35	Y
Cabbage	0	98-100	R, F	90-180	Y
Cantaloupe	3 to 5	95	H, F	10-14	Y
Carrots, topped	0	98-100	I, R	28-180	Y
Cauliflower	0	90-98	H, V	20-30	
Celery	0	98-100	I	14-28	Y
Cherries, sweet	-1 to -0.5	90-95	H, F	14-21	
Corn, sweet	0	95-98	H, I, V	4-6	
Cranberries	2 to 4	90-95		60-120	
Cucumbers	10 to 13	95	F, H	10-14	Y
Eggplant	7 to 13	90-95	R, F	10-14	Y
Endive	0	90-95	H, I	14-21	Y
Garlic	0 to 1	65-75	N	90-210	
Grapefruit	10 to 16	85-90		28-42	
Grapes	-0.5 to 0	85	F	56-180	
Kiwifruit	0	95-100		28-84	Y
Leeks	0	95-100	H, I	60-90	Y
Lemons	10 to 13	85-90		30-180	

Contd...

Table 4.2–Contd...

Fruit/Vegetable	*Temperature (°C)*	*Relative Humidity, %*	*Pre-cooling Method*	*Storage Life, Days*	*Ethylene Sensitive*
Lettuce	0	85-90	H, I	14-21	Y
Limes	9 to 10	85-90		21-35	
Mushrooms	0	95		12-17	
Nectarines	-0.5 to 0	95	F, H	14-18	Y
Okra	7 to 10	90-95		7-14	
Onions, bulb	0	65-70	N	30-180	
Onions, green	0	95-100	H, I	7-10	
Oranges	0 to 9	85-90		21-56	
Peaches	-0.5 to 0	90-95	F, H	14-28	Y
Pears	0	90-95	F, R, H	60-90	Y
Peas, in pods	0	95-98	F, H, I	7-10	Y
Peppers, bell	4 to 13	90-95	R, F	12-18	Y
Peppers, hot	7 to 10	60-70	R, F	14-21	Y
Pineapple	7 to 13	85-90		14-36	
Plums	0	90-95	F, H	14-28	Y
Potatoes, early	10 to 16	90	R, F	56-140	
Potatoes, late	4 to 10	90	R, F	56-140	Y
Pumpkins	10 to 16	50-75	N	84-160	
Raspberries	0	90-95	R, F	2-3	Y
Rutabagas	0	98-100	R	120-180	
Spinanch	0	95-100	H, I	10-14	Y
Squash, summer	4 to 10	95	R, F	7-14	Y
Squash, winter	10 to 13	50-70	N	84-150	
Strawberries	0	90-95	R, F	5-10	
Sweet potatoes	13 to 16	85-90	N	120-210	Y
Tangerines	4	90-95		14-28	
Tomatoes	17 to 20	90-95	R, F	7-28	Y
Turnips	0	95	R, H, V, I	120-150	
Watermelon	10 to 16	90	N	14-21	

*F: Forced-air cooling; H: Hydro-cooling; I: Package icing: R: Room-cooling; V: Vacuum-cooling; N: No pre-cooling needed.

Source: USDA Agricultural Marketing Service, Kansas State University Cooperative Extension Service.

Chilling Injury

Many vegetables and fruits store the best at temperatures just above freezing, while others are injured by low temperatures and will store best at 7 to 13°C. Both time and temperature are involved in chilling injury. Damage may occur in a short time if temperatures are considerably below the danger threshold, but some crops can withstand temperatures a few degrees into the danger zone for a longer time. The effects of chilling injury are cumulative in some crops. Low temperatures in transit, or even in the field shortly before harvest, add to the total effects of chilling that might occur in storage.

Crops such as basil, cucumbers, eggplants, pumpkins, summer squash, okra, and sweet potatoes are highly sensitive to chilling injury. Moderately sensitive crops are snap beans, muskmelons, peppers, winter squash, tomatoes, and watermelons. These crops may look sound when removed from low temperature storage, but after a few days of warmer temperatures, chilling symptoms become evident such as pitting or other skin blemishes, internal discoloration, or failure to ripen. Tomatoes, squash, and peppers that have been over-chilled may be particularly susceptible to decay.

Preventing Moisture Loss

While temperature is the primary concern in the storage of fruits and vegetables, relative humidity is also important. The relative humidity of the storage unit directly influences water loss in produce. Water loss can severely degrade quality. For instance, wilted greens may require excessive trimming, and grapes may shatter loose from clusters if their stems dry out. Water loss means saleable weight loss and reduced profit.

Most fruit and vegetable crops retain better quality at high relative humidity (80 to 95 per cent), but at this humidity, disease growth is encouraged. The cool temperatures in storage rooms help to reduce disease growth, but sanitation and other preventative methods are also required. Maintaining high relative humidity in storage is complicated by the fact that refrigeration removes moisture. Humidification devices such as spinning disc aspirators may be used. Even buckets of water will increase humidity as the fans blow air across the water's surface and increase evaporation. Keeping the floor wet is helpful, though messy and potentially hazardous to two-legged creatures. Frequent cleansing with a weak chlorine solution will be needed to prevent harbouring of disease organisms in water and produce scraps on the floor. Crops that can tolerate direct contact with water may be sprinkled to promote high relative humidity.

Sanitation

Sanitation is of great concern to produce handlers, not only to protect produce against postharvest diseases, but also to protect consumers from food borne illnesses. Use of a disinfectant in wash water can help to prevent both postharvest diseases and food borne diseases.

Ethylene

Ethylene, a natural hormone produced by some fruits as they ripen, promotes additional ripening of produce exposed to it. The old adage that one bad apple spoils the whole bushel is true. Damaged or diseased apples produce high levels of ethylene and stimulate the other apples to ripen too quickly. As the fruits ripen, they become more susceptible to diseases. Ethylene producers should not be stored with fruits, vegetables, or flowers that are sensitive to it. The result could be loss of quality, reduced shelf-life, and specific symptoms of injury. Ethylene producers include apples, apricots, avocados, ripening bananas, cantaloupes, honeydew melons, ripe kiwifruit, nectarines, papayas, passion-fruit, peaches, pears, persimons, plantains, plums, prunes, quinces, and tomatoes.

Thus postharvest handling is the final stage in the process of producing high quality fresh produce. Being able to maintain a level of freshness from the field to the dinner table presents many challenges. A grower, who can meet these challenges, will be able to expand his or her marketing opportunities and be better able to compete in the market. Improved postharvest management depends on the quality and efficiency of handling, processing and preservation techniques used. Increase in agricultural production could be nullified because of inappropriate and unreliable postharvest management employed. In nut shell, postharvest management is focussed on reducing postharvest losses by devising better storage and processing technologies and improving the transit of goods from producers to the consumers.

References

Bachmann, J. and Earles, Richard 2000. Postharvest handling of fruits and vegetables. ATTRA-National Sustainable Agriculture Information Service, California. Accessed at http://attra.ncat.org/attra_pub/PDF/postharvest.pdf.

Golob, P., Farrell,G. and Orchard, J.E. 2002. Crop Postharvest: Science and Technology. Volume 1:Principles and Practices. Blackwell Publishing Company, USA.

Satish, H.S. 2000. Packaging of fresh fruits and vegetables-International Practices. In: Modern Food Packaging, Indian Institute of Packaging, Mumbai.

Singh, Sanjay 2002. Technology Information, Forecasting and Assessment Council study reports on Packaging.Technology Bhawan, New Mehrauli Road, New Delhi-16. Accessed at Internet.

Chapter 5

Prospects of Postharvest Technology and Value Addition in Pulses

R.K. Goyal and S.M. Ilyas

Central Institute of Postharvest Engineering and Technology, PAU Campus, Ludhiana – 141 001 (Punjab)

Grain legumes or pulses occupy an important place in the world food and nutrition economy. These are important constituents in the diet of a very large number of people, especially in the developing countries and are good sources of protein, which help to supplement cereal diets, improving their protein nutritive value. These also provide substantial quantities of minerals and vitamins in the diet. Although most legumes are consumed as dry grains, immature green pods or green seeds are also used as vegetables.

The availability of grain legumes over the last a few years has dropped because their production has not been very profitable as compared with other crops. The consumer demand for legumes has, however, not fallen and prices have increased considerably. Improved conservation and processing to reduce postharvest losses and the manufacture of economically priced products based on grain legumes would help to increase the supplies. The development of this industry would provide additional rural employment, improve nutritional standards, offer a better price to the grower and ensure quality supplies at lower prices to the consumer.

In Asia and Africa, a substantial protein of the legumes is consumed after having been milled (affecting removal of the husk and splitting) or after some of processing. However, most of the commercial technologies available for this purpose are either obsolete or inadequate and result in heavy losses due to breakage and powdering of the grain. Some successful efforts have been made to develop improved technologies to reduce milling losses and improve product quality. Similarly, there is a need for development and utilisation of improved technologies for the manufacture of products based on grain legumes. The reduction of pollution, recovery of cotyledon material from the waste and improved utilisation of energy are some of important areas of concern.

Milling of Pulses

Generally pulses are consumed in India after dehulling and splitting and converting them into "*dal*". Removal of the seed coat (husk) reduces roughage, improves storability and palatability for consumption in various forms. It also improves cooking quality and digestibility. Conversion of pulses into "*dal*" is the third largest food processing industry after rice and wheat milling industries. It is estimated that about 75 per cent of the pulses produced in the country are converted to *dal*. Milling of pulses has been practiced as a tiny and small-scale rural operation from the times immemorial but lately it has been converted into a fairly large commercial operation (Kurien and Parpia, 1968). One estimate indicates that about 30 per cent of the production of pulses is still carried out on a cottage scale level by the farmers in rural areas using traditional techniques.

There are several small-scale millers who process 0.1-0.5 tonne per day either by custom milling or for trade purposes as a home-scale operation. However, with the advent of organized milling systems, most of these processing systems are disappearing as the product is inferior in quality and yield. For this reason, the *dal* available in the market mostly comes from the large-scale commercial mills. There are about 7,000 *dal* mills working in various parts of the country processing different pulses throughout the year.

Traditional and Conventional Methods

Pulses go through several primary processes juice scouring hulling (dehusking), roasting, puffing, grinding, splitting, etc. before they can be used in different food preparations. Hulling, practised widely in Asia and Africa either on a home-scale or as a cottage industry, produces refined cotyledons with good appearance, texture, and cooking qualities. Dehusked grains are easily digested and efficiently utilised by the body. As the husk tightly envelops the endosperm, usually through a thin layer of gums and mucilage, the primary step in hulling involves laborious procedures.

The oldest and most common home-scale technique for hulling grain legumes is to pound them in a mortar with a pestle, either after spreading the grains in the sun for a few hours or after mixing them with a little water. The husk is then winnowed off to get the clean cotyledons. Methods followed in the home or village industry or in commercial mills are usually similar in principle, but different in the use of technique for better yield, operational efficiency, and large-scale application. Home-scale hulling consists generally of two steps: (i) loosening of the husk by wet or dry methods and (ii) removal of the husk and cleaning. In South Asia, the first step is achieved by sun-drying the raw mature grains as such, or after they have been treated with oil and/or water. In some areas, grain is steeped in water for two to eight hours prior to sun drying. Grain varieties whose husks are tightly attached to the cotyledons are soaked and then treated with red-earth paste before being sun-dried. The steeping technique to loosen the husk is also practised in several Southeast Asian and African countries.

Pounding the grain in a mortar with a pestle or grinding in a hand-operated wooden/stone sheller accomplishes dry-method husking. Winnowing thereafter separates the husk. This is a common practice where dry cotyledons or grain-legume flours are used in food preparations. When a batter or dough is to be prepared, the soaked grain is either rubbed by hand to remove the husk, after which it is separated by flotation or wet-ground in a stone grinder. In several Southeast Asian countries, this technique is used for extracting starch from mung beans, the husk being removed by straining dilute slurry through coarse muslin or other cloth. Home-scale methods are generally laborious and not always hygienic. In village industries, the techniques employed for loosening the husk are:

1. Prolonged sun-drying until the husk is loosened
2. Application of small quantities of oil, followed by several hours or even days of sun-drying and tempering

3. Soaking in water for several hours, followed by coating with red-earth slurry and sun-drying
4. Soaking in water for several hours to loosen the husk before manufacture of food products and
5. A combination of these techniques.

There are no standard procedures developed for any specific variety of pulse, which really calls for characterization of pulse varieties in terms of physical and thermal properties and their processability–a fact which is often lost sight of by the breeders.

Removal of the loosened husks from the grain in the dry-milling technique is commonly done in small machines. Hand or power-operated under-runner disc-shellers or grinders with emery or stone contact surfaces are used. A plate mill with a blunt contact surface is sometimes used both to husk and split the soaked and dried grains. After aspirating or winnowing off the husk, sieving separates the split cotyledons. Remaining unsplit whole grains are similarly processed until almost all the grains are husked. In certain parts of India, oil-treated and sun-dried grains are husked in an Engelberg-type rice-huller after being mixed with 2-3 percent stone powder. Sound kernels are removed by sieving, while the husk, powder, and small brokers remain in the stone powder.

Hulling methods are not one-step operations. About 50 percent removal is achieved in the first operation. After separation of the husked split-cotyledons (*dal*), the process is repeated several times until almost all the grains are converted into *dal*. In the process, excessive breakage with powdering of grains occurs because of repeated splitting and husking operations. Complete removal of the husk from the grain is not always achieved in cottage-scale processing, particularly with some varieties and hence unpolished (split and unhusked) *dal* is sometimes put on the market. The cooking quality of *dal* prepared by wet methods is usually poor. This is especially true of the pigeon pea for which cooking time increases with the duration of soaking. However, such *dal* has an attractive appearance and more desirable processing

Hulling of legumes on a commercial scale is generally based on dry-processing techniques. Many of the operations, particularly husking and splitting, are mechanised. The drying is done in large yards and is completely dependent on sunshine. Legumes such as pigeon pea, black gram, and mung bean, which are more difficult to husk, require more of the oil or water treatments followed by prolonged sun-drying (pre-milling treatments), while grains such as chickpea, lentil, pea etc., with more easily removable husks, require short periods of sun-drying and fewer oil or water treatments. Sometimes these grains are given an initial "pitting" in the roller mill to crack the husk and improve the absorption of oil or water. In the case of black gram, the coating of wax and dust is removed by initial scouring in a roller mill, facilitating absorption of water or oil. Husking and splitting are done either in a single operation or, more advantageously, as independent operations. Moisture addition adversely affects husking, but it helps to split the grain. Addition of water prior to husking helps to induce simultaneous splitting, but this often leaves patches of husk that have to be removed by scouring in polishing machines. As the husk forms 10-16 percent of the grain, a maximum theoretical yield of 84-90 percent of kernels should be possible during hulling. In practice, the yields vary from 68 to 76 percent as a result of breakage, powdering, and other milling losses. In wet methods, water-soluble nutrients are also lost in the soak-water.

Improved Technologies for Hulling of Grain Legumes

The improved method developed at the Central Food Technological Research Institute, Mysore for hulling grain legumes commonly consumed as *dal* or split-grain legumes in India, Pakistan, Bangladesh, Nepal and some other countries involves a conditioning technique of moisture adjustment of the grain to a critical level in order to loosen the husk. In this process, the grain is exposed to heated air, at a specific temperature appropriate for each variety, for a predetermined time and equilibrated to the critical moisture level with gradual aeration in tempering bins. The husk is removed in an improved abrasion-type hulling machine that accomplishes almost complete removal in a single pass with the least possible scouring or breakage of the endosperm. The whole husked grain can be split in a splitting machine after suitable conditioning for which the technology and equipment have already been developed. The method is independent of the influences of climate and variations among the varieties. This improved technology has been shown to increase the yield by 5-10 per cent. The time taken for processing is also less and the cost of operation is lower. Small commercial models of complete automatic milling plants of 0.5-2 tonne per hour capacity have been designed and put into operation. The process, originally developed for pigeon pea has been adapted with suitable modifications for processing other legumes such as chick pea, mung bean, black gram, lentil, pea, soybean and some other beans.

In the Prairie Regional Laboratory, Saskatoon, Canada, a hill-threshing unit consisting of carborandum discs has been successfully used for dehulling cowpea. Initial findings indicate that the unit is capable of mechanically hulling brown Nigerian cowpea that can subsequently be converted to flour. The husk is removed by the abrasive action of rotating discs, the amount removed being dependent upon the through-put and retention time. An acceptable product would necessitate removal of about 27 percent as polish from the cowpea, whose husk content is 3-5 percent (22-25 per cent loss of kernel). This unit is being put into rural-scale operation for processing cowpea in some African countries.

The hulling characteristics of grain legumes are influenced by the variety, season, when harvested, and agro-climatic situation. Larger of bold-grain varieties are easier to hull, give a higher yield and are preferred by millers, while the smaller varieties require repeated and severe pre-hulling treatments and complex procedures. Freshly harvested grain and winter crops are more difficult to process, possibly because of their higher moisture content. Such grain is either stored for some time to reduce moisture or treated with limewater or a solution of sodium carbonate to loosen the husk.

Whole legumes or husked splits are either ground dry into a flour or ground wet into a batter for a number of sweet and savoury preparations, either alone or in combination with cereals and millets. The eating quality of many of these products, particularly the texture, depends on the composition of the flour, degree of fineness of grinding relative proportion of particles of different mesh grades and cooking conditions. Chickpea, pea, black gram and cowpea are the common grain legumes ground wet or dry.

Different Pre-milling Treatments

Wet Method

This method of dehulling generally involves water soaking and sun-drying and is common in many parts of India (Singh and Jambunathan, 1981). The wet-method has the advantage of facilitating good dehusking and splitting of the cotyledons, giving less breakage though it may adversely affect the cooking quality. The method is also labour intensive and is completely dependent upon climatic conditions for drying the soaked seeds. The entire process usually takes about five to six days and only limited quantities can be processed at any given time. *Dal* produced by wet-method tastes better but takes longer time to cook. The soaking of pigeon pea in water for the time ranging from 2 to 14 hours is a common practice in Maharashtra, Uttar Pradesh and Madhya Pradesh. Soaking for longer periods is preferred in certain region where pulses are processed in summer season (Singh, 1994). Soaking is generally practised at village level. In large scale operation spraying the seeds with water increases the seed moisture level. Comparatively smaller dehusking units practise alternate wetting and drying known as conditioning (Saxena, 1985). Dehulling can be rendered easier by prolonged soaking in water for 12h or more but *dal* remains uncooked and tough even on prolonged boiling.

Dry Method

The cleaned and graded grains are passed through an emery-coated roller for initial pitting or scratching of the husk. Pitted grains are thoroughly mixed with about 1 per cent oil and spread in a thick layer for sun drying in the drying yards for 2-5 days. Grains are heaped during the night to preserve the heat. About 2-3 per cent of water is added to the grains before milling and the grains are subsequently passed through the roller for dehusking by abrasion. In this process, about 40-50 per cent of the grains are de-husked and major portion of these gets split simultaneously. The loosening process may be slow but the husk can be totally loosened if the treatment is extended to several days. This method is said to produce *dal* that cooks faster than the *dal* produced by the wet-method. The major disadvantage of this method is high dehulling losses due to breakage and powdering.

Sodium Bicarbonate Treatment

Krishnamurthy *et al.* (1972) tried the use of sodium bicarbonate, sodium carbonate, sodium hydroxide, acetic acid and ammonia as a replacement for vegetable oil in the traditional process and reported a considerable improvement in *dal* yield when sodium bicarbonate was used. Reddy (1981) used sodium bicarbonate (5 per cent solution) and reported a *dal* yield of 75 per cent. Saxena *et al.* (1981) treated pigeon pea grain with aqueous solution of calcium hydroxide, sodium carbonate and sodium chloride of different normalities. Sodium bicarbonate solution of 1N was reported to be the most effective, resulting in *dal* yield of 78 per cent in addition to loosening of husk. It also reduced the cooking time of the resulting *dal*.

Srivastava *et al.* (1988) showed that the pigeon-pea grain soaked in 6 per cent sodium bicarbonate gave improved dehusking efficiency They however observed an increasing loss of crude protein due to leaching as the concentration of sodium bicarbonate solution was increased. They reported the total of crude protein in the range of 6-8 per cent at 6 per cent $NaHCO_3$ concentration.

Mayande (1987) reported a maximum hulling efficiency of 90 per cent for grains treated with 10 per cent $NaHCO_3$ solution at grain to solution ratio of 20:1 for 5 hours. Broken percentage was the highest in control and was comparatively less in grain to solution ratio of 20:1. The powder was the minimum in control and did not show any specific trend due to the effect of time or ratio.

Enzyme Treatment

Verma *et al.* (1993) employed edible enzyme as a husk-loosening agent. He reported a maximum hulling efficiency of 88.93 per cent at an enzyme concentration of 0.08g protein per 260g pigeon-pea grain. Grains were treated with the enzyme and allowed to incubate. During this period of incubation, enzymatic hydrolysis took place which brought about the biodegradation of complex molecules of the grain. The complex gums were degraded which resulted in easy dehusking. It established that a lesser force was required to bring about the dehusking of enzyme treated grain. The action of the enzyme also disturbed the microstructure of the grain affecting its strength. He further reported an increase in the protein digestibility and 37.03 per cent reduction in cooking time. Further, this *dal* was reported to cause less flatulence due to fermentation which broke down the poly-saccharides responsible for causing flatulence, in many people.

Advance *dal* Milling Technologies

Processing

A large number of varieties of pulses are grown, processed and consumed in India such as pigeon pea, mung-bean, urd-bean, chickpea, lentil, pea, etc. The requirements of processing differ not only as per the variety but also often differ for the same variety of pulse grown and harvested in the areas having different soil, climatic conditions. Almost all the pulses have to undergo the processing sequences as shown in the flow chart (Figure 5.1).

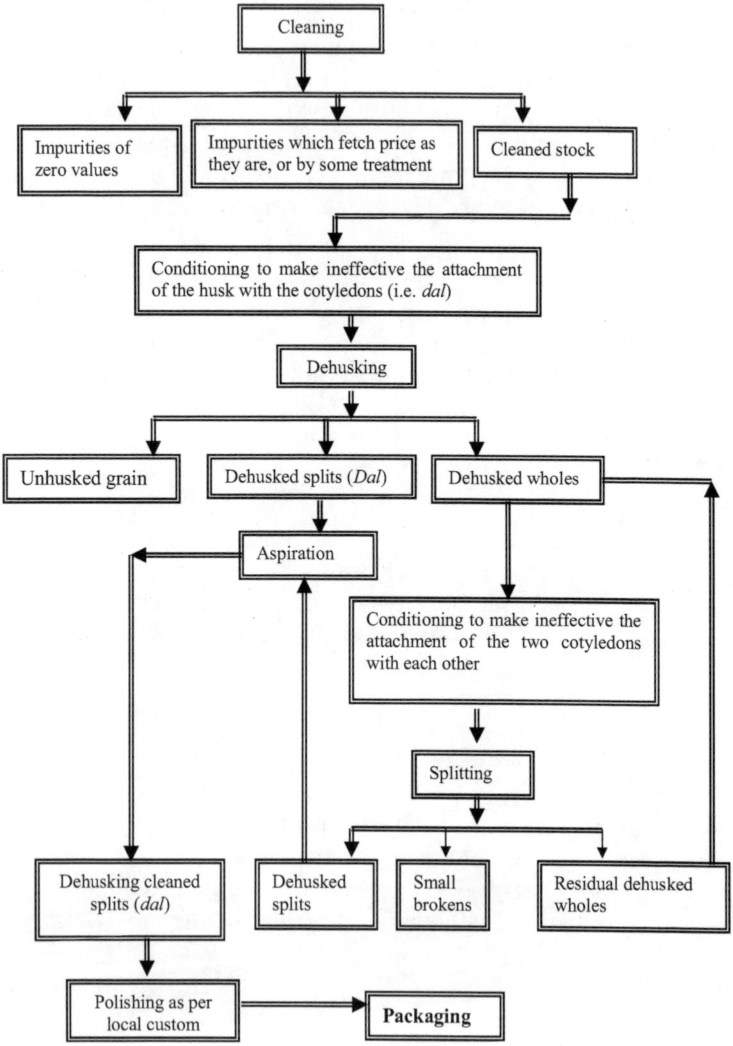

Figure 5.1: Sequences in Processing of Pulses (*Source*: Matanhelia, 1980)

Cleaning and Size-grading

Cleaning

Elimination of all impurities, foreign matters, a damaged and/ or unwanted grain is the first step in the processing of pulses.

Aspiration System

Centrifugal fans with low pressure chambers, cyclones and air filters are well in use to draw the dust, chaff and other light materials from the stocks- in-process, and to deliver the impurities thus drawn off separately, so as to suit the processing requirements. The junction box which is also a low pressure box sets aside the particles and the heavier materials. The cyclone dust-collector collects and sets aside the lighter impurities and dust, etc. The air discharged from the cyclone is however not completely dust- free and may contain dust upto 0.5 per cent to 1 per cent. If such dust is the fine pulse powder which can be sold for some price, it has to be recovered from the air before its discharge into the atmosphere. Sometimes the state laws make it compulsory that the air should be completely dust-free before being discharged into the atmosphere.

Sieving

Out of all the sieving machines, a reel machine seems to be the best equipment for the purpose of setting aside the impurities according to size. The *dal* mills in India are using crude types of reel machines for more than the last fifty years. Since improved machines are now available, they are preferred. A cylinder with suitable numbers and specifications of perforated sets and supported by two suitable bearings rotates slowly inside the body of the machine having suitable number of partition chambers, collecting hoppers and inspection doors. While the stocks enter the cylinder and travel from the point of entry to the exit, the impurities and the size graded stocks drop down in the collection hoppers according to size from where they are moved off as per the processing or disposal requirements. The rotation of the cylinder at slow speed accounts for low maintenance cost and interruption–free working. Provision of scrappers keeps the perforations clean and ensures high efficiency continuously. The reciprocating sieve decks are less preferred because the repair/maintenance is difficult and costlier.

Size grading

Size grading of the stocks in process is essential on account of the reasons such as:

1. A size graded stock offers better efficiency at the gravity tables or the destoners.
2. Flowering and seed formation of the pulse takes place in the field during a prolonged period. The seeds grown in the later periods are less mature and differ in characteristics requiring different adjustments at the conditioning stage, and
3. Product of the graded raw material has better appearance and fetches better price.

Destoners

The stocks-in-process carry with it impurities such as stones, hardened mud-particles, metal and glass pieces. These impurities when equal in size to the stocks to be processed do not separate by sieving. However since these are of higher specific gravity, a destoner can conveniently separate them.

Wet destoners and dry destoners are available and the latter is preferable. Pulses being hygroscopic in nature attract lot of moisture, even to unwanted extent, when the stocks are passed through water in a wet destoner and the miller does not have a control on it. When a dry destoner is used, water only as per the conditioning requirements can be added in a paddle screwworm. Gravity tables can also be used with advantage but they are more costly and intricate.

Pneumatic Separators

Even after cleaning aspiration, screening and destoning, the stocks-in-process may contain some damaged, weevilled, under-developed or shriveled grains which are lighter in weight (specific gravity) than the sound grains. They can conveniently be separated by pneumatic separators.

Cockles Cylinders or Carter Discs

These are used to separate grains which do not differ in size or gravity but differ in shape. Presence of such unwanted grains is particularly significant in the case of lentil. Their separation, when done by sieving, removes grains alongwith which can otherwise be converted into dehusked wholes or dehusked splits to be produced as per the marketing requirement. These impurities can conveniently be separated by cockles cylinder machines. The unwanted grains are generally round in shape while the lentils are flat. Spiral towers can also be used for separation.

Magnetic Separators

Permanent magnets of good standard quality should be used to separate the iron particles from the stocks-in-process at the points of their entry into the pitting machines, de-huskers or the shellers otherwise these will badly damage the emery rolls or the emery disc and may even cause fire.

Conditioning

Conditioning loosens the binding of the husk with the cotyledons and the binding of the two cotyledons with each other. Conditioning requirements differ not only from pulse to pulse but also for the same variety grown in different areas depending upon the type of soil, climate and harvesting season. Three alternative systems of conditioning, dehusking and splitting are being suggested (Figures 5.2a, 2b and 2c). The de-husking and splitting of the heated stocks take place simultaneously in a sheller (fitted with emery disc of the improved design and coated with emery or carborandum of the required granulites), when the material reaches the point of contact at the circumference of the discs. The stocks after being heated in the sheller go for screening whereas the de-husked splits are at the heated temperature and have to be cooled before packing.

Type-A conditioning system is recommended for pigeon pea, mung-bean and urd-bean and alike varieties of pulses which are most difficult to de-husk and split. These require more intensified conditioning effect to loosen the bond of the husk to the cotyledons and the bond between the two cotyledons. Heating and tempering are provided twice to get the desired effects. Type-B conditioning system which provides a single heating and single tempering system is suitable for de-husking and splitting of dun pea, faba beans, broad bean, chickpea, soybean and alike varieties of pulses. Conditioning system of type-C is especially suitable for de-husking and splitting of chickpea. An over treatment to some of the grains can damage

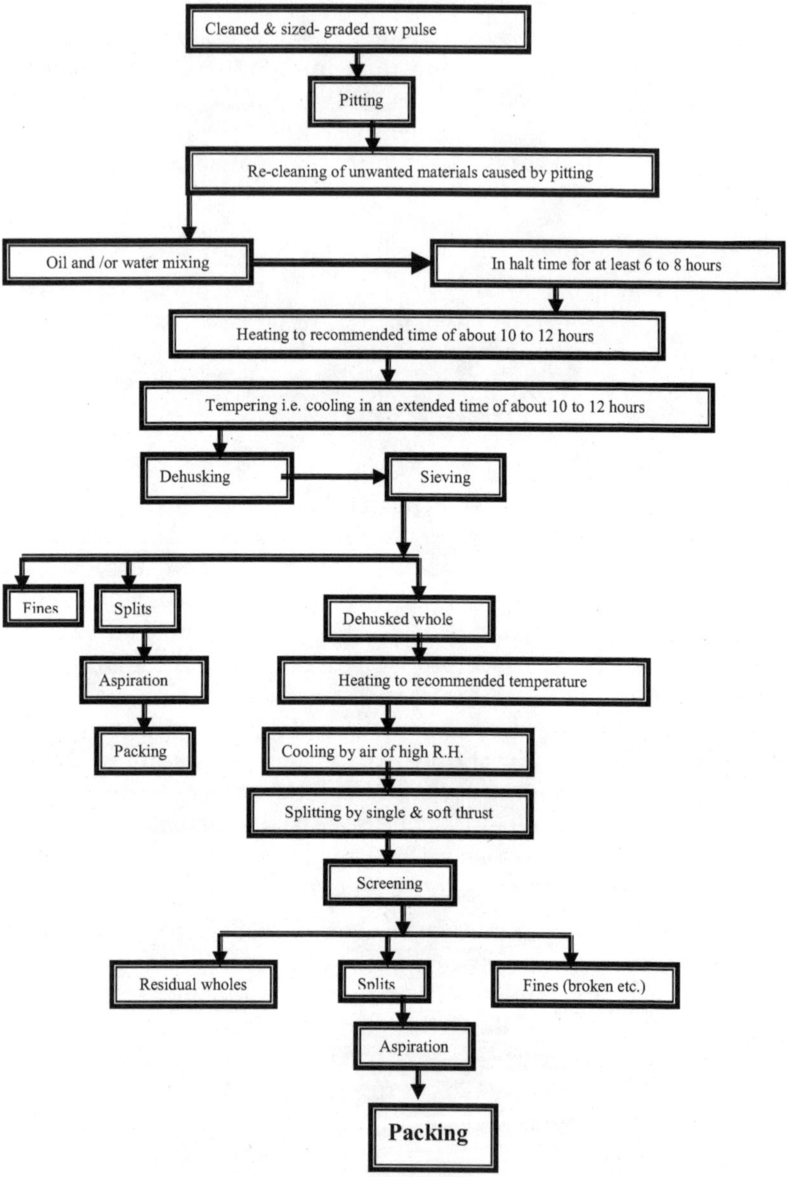

Figure 5.2(a): Conditioning System Types 'A'
(*Source*: Matanhelia, 1980)

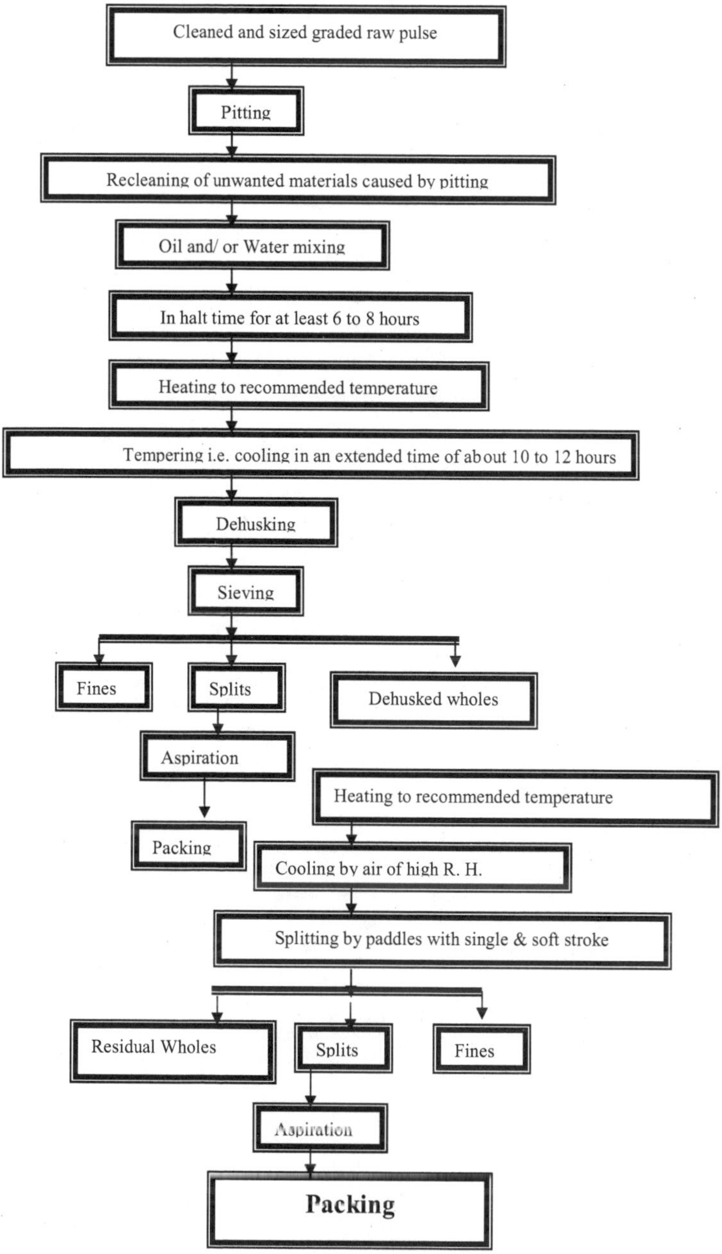

Figure 5.2(b): Conditioning System Type 'B'
(*Source*: Matenhelia, 1980)

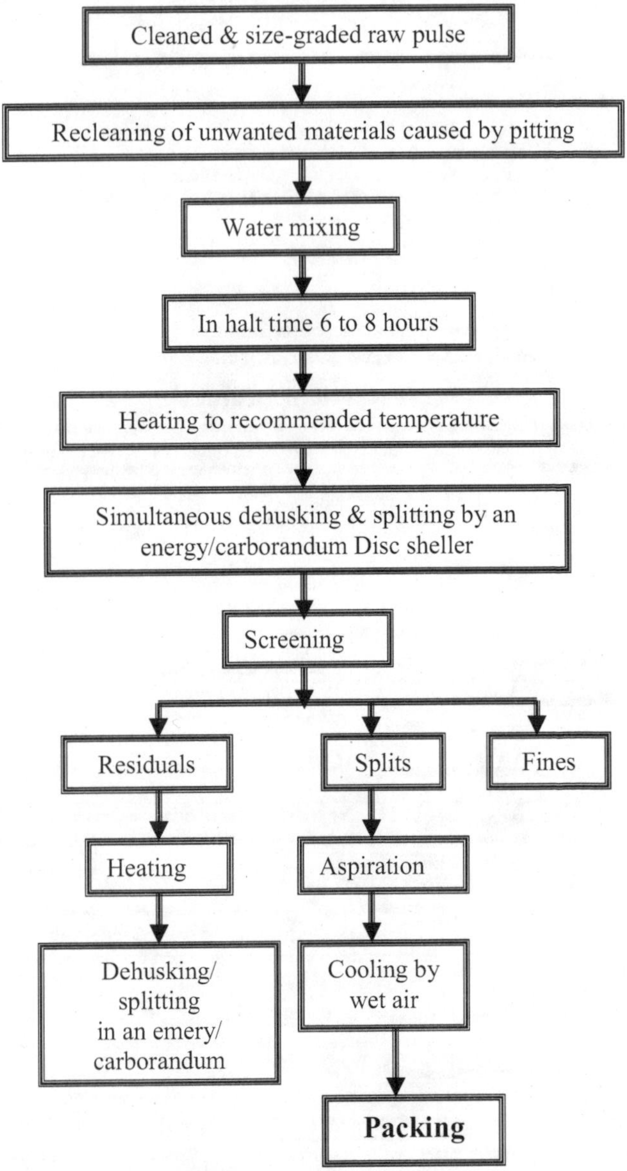

Figure 5.2(c): Conditioning System Type 'C'
(*Source*: Matenhelia, 1980)

the quality, while under- treatment may result in husk- patches on dehusked splits.

Pitting

Pitting is done to cause some cracks and pores in the seed coat to enable better penetration of the various conditioning effects such as those given during oil and/or water mixing, heating and tempering, etc. The ideal pitting would be such that the cracks or the pores may not be visible by naked eye. De-husking or splitting is to be avoided as far as possible at this stage although some amount of dehusking and/or splitting is quite unavoidable on account of the condition effects.

Oil/water Mixing

Mixing of oil and/or water followed by an in-halt time of 6 to 8 hours sometimes produces very good effect for the better conditioning of the stocks-in- process and this can be well adopted to supplement the effect of conditioning at subsequent stages whenever necessary. While the mixing of water helps de-husking, it also leads to excessive splitting and is sometimes avoided for specific product requirement.

Heating and Tempering

It has very well been established that the heating of almost all the pulses to the recommended temperature followed by tempering i. e, cooling in extended time, has a very positive effect in loosing the binding of the husk to the cotyledons. This process of heating is done once or twice depending on the intensity of conditioning required. The temperature of the air used for heating as well as the temperature, to which the stock is heated, will vary from pulse to pulse. Similarly, it has been observed that allowing more time for cooling of the heated stock during tempering has a better effect but the equipment-cost as well as the cost of the material which will remain in process, puts a limit for the time which can be allowed at this stage. The equipment is accordingly designed. Air is preferred as the media for carrying the heat from the source of heating to the grains for heating and also from the grains to the atmosphere for tempering *i.e.,* cooling. It also helps in bringing down the moisture to the required level.

For heating of the air, electrical air- heaters or fuel-oil burners using furnace oil, H.S.D. or L.S.D oils or the LPG have all been very successfully tried. In the case of heating of the air by fuel-oil burners, the heating system should be indirect so that the gases do not injuriously affect the material.

De-husking

Emery/carborandum coated rolls or cones placed and operated in vertical and horizontal position have been tried with varying results. The rolls and cones placed vertically have been observed to be totally unsuitable. The material from the feed point drops instantaneously to the lower part where it forms practically a jam. While there is too much overloading and power consumption, not all the grains get the proper contact with the abrasive surface of the emery roll. It also causes too much powdering and the dehusking is poor. On the other hand, emery/carborandum coated rolls of conical shape and horizontally placed give the best results. Almost all the grains get dehusked. There is no rubbing or powdering or quibbling. The quality of the product is naturally better and the yield higher. An emery roll-coated with the most appropriate grains of emery and/or carborandum and accurately balanced, rotates inside a shell of which the clearance to the roll can be adjusted while the machine is in operation to suit the specific requirements. The roll is fixed to a shaft provided with ball bearing of the best quality suitable to withstand the working loads. The shell is fitted with covers of mild steel and perforated sheets of sufficient thickness. The unit is placed in a body fabricated from steel and provided with covers and inspection doors. A powerful magnet is provided at the inlet to eliminate the entry of iron pieces in the machine along with the stocks to be treated so as to avoid the possibility of damage to the roller or shell cover. The inlet has also arrangement to regulate the feed of the stocks to the machine as well as for instantaneous shut-off. The delivery is provided with mechanism to regulate the pressure inside the shell. All these arrangements enable best dehusking results.

For splitting after making the binding of the two cotyledons ineffective by the conditioning system, the grains just need a single thrust and not harsh multiple thrusts to split. There is no quibbling and this again increases the yield.

Husk Aspiration System

The de-husked splits received by screening after dehusking as well as after splitting contain husk of pretty big size and are therefore to be cleaned by an efficient system of aspiration. There being a little difference between the weight of the husk and the splits, the system requires provisions for every minute adjustment so that no splits are carried away by the air with husk and at the same time no husk should remain with the splits. The new type of aspiration equipment are well designed to adjust the air to the required amount and pressure and when the same is passed at cross currents with the flow of the stocks to be treated, it efficiently draws off the husk and lets the cleaned dehusked splits to the packing. The equipment comprises of (i) cascade where the strokes meet at cross current air at required pressure and speed; (ii) junction box, which is also a low pressure box to recover the small quibbling which are carried away by the air; (iii) cyclone dust collector; (iv) air locks and (v) air filters, etc. all suitably arranged. For still finer separations, vibro-aspirators are recommended instead of the cascades.

Prospects

Modernisation of *dal* Milling Industry

Traditional technologies for making *dal* are laborious, time-consuming and completely dependent on climate and low yielding. As such there is a need to modernise the processing steps and machinery involved for better and quick recovery to reduce the cost of processing and make the process independent of climatic conditions. In modernising the *dal*-milling industry, the CFTRI, Mysore, has employed the "conditioning technique" to loosen the husk without resorting to sun-drying and oil and water application. This step has been mechanised with the introduction of conditioning units where the grains are incipiently roasted by a counter-current through flow technique and tempered in specially designed tempering bins where moisture is reduced to the desired levels. The husk thus loosened is then removed in an abrasion-type dehusking machine wherein almost complete removal of the husk is achieved in a single operation. The scouring and breakage losses are minimised. As the milling characteristics and physical properties of pulses vary, appropriate processing conditions and steps have been developed with the "conditioning technique" which is almost

5-10 units more than that obtained by using the traditional methods. The resulting products are also better in cooking quality.

Pulses in Snack Foods

Snack foods prepared by heating, toasting (parching) or pulsing of pulses are traditional household items consumed with cereals, malt and oilseed meals. Pulsing is done to impart a light and porous structure in addition to improving the aroma and flavour. Chickpea is the commonly pulsed pulse, although pea and cowpea are also pulsed. Technique of pulsing in home-scale processing involves drying of chickpea by beating with sand at low temperatures, soaking the grain in water for a few seconds, roasting again with sand heated at 200°–250° C and then beating in cloth bags. Recently the drying and beating technique has been mechanised with the introduction of "gram frying machines".

Development of Improved Technology

Batch-process which presently involves excessive material handling can be reduced by introducing continuous processing system. The *dal* yield varies from one milling unit to another depending upon the equipment and processing methods used. Thus there is a need for development of improved method of milling where millers can get cleaned and graded raw materials which will reduce processing cost and material handling.

Utilisation of Existing Rice Milling Machinery for Pulse Milling

Abrasion rice polisher, under-run disc (URD) sheller and centrifugal sheller are traditionally used for pulse milling. Under-run disc sheller and centrifugal sheller are being extensively used in small-scale mills. Abrasive rice polisher (Satake make) has been tried on laboratory scale only. Therefore, it can be directly useful for milling of pulses when it is scaled up and evaluated on large-scale basis. Basically, URD sheller with a different coating is used for dehusking and splitting. Rice polisher unit attached with a blower can be used for polishing.

Development of Pre-milling Treatment Processes

The pre-milling treatments (to loosen the husk or to reduce the binding between the cotyledons) should be such as to meet the exacting requirements of the properties of grain so that its processing could be done economically with newer varieties of pulses. The use of oil for loosening the husk during milling should be eliminated to save more than 15000 tonnes of edible oil used annually in India in this operation alone. The suggested pre-milling treatments need to be verified on a large-scale for different pulses before giving recommendations.

Development of Equipment for Cleaning, Grading, Drying and De-husking

Milling quality is not improved by pre-milling treatments alone but the use of proper machine is desirable to further improve upon recovery rates. Normally, two types of cleaners are available *viz.* rotary-screen-cleaner for raw material cleaning and reciprocating screen for *dal* grading which are employed by some manufacturers. The size of perforations varies as per the type of grain to be processed. However the sieves are to be designed and fabricated according to the geometry of different pulse crops.

The traditional method of sun-drying is very time consuming, weather-dependent and labour intensive process. Taking into consideration all these aspects, a *dal* dryer, which could be run on the by-products of the industry, would be quite helpful. The cost of the dryer should be within the reach of small millers. Another advantage with mechanical drying is that the large drying yards possessed by the mills can be put to some more productive commercial use.

Although much information is available on grain drying which could be adopted for pulse drying, the drying characteristics of different pulses need to be studied for designing proper conditioning parameters for each pulse crop. It is reported that red gram should be dried at 55 to 60° C and upto 6-6.5 per cent moisture content for better dehusking. Several research workers have also derived empirical drying equations.

An improved roller machine (abrasion principle) for efficient de-husking of grains without causing much loss may be developed. Design and structural features of abrasion type and attrition type machines for different operations like dehusking, splitting, polishing etc. may be identified. Some millers use vertical emery burr mill whereas others may use concentric cylinder roller mill for dehusking and splitting. Hence there is a need to standardise the equipment used in milling of pulses. This would lead towards reducing cost of fabrication of equipment and at the same time availability of proper spare parts can be ascertained.

Development of De-husked Whole Pulse (*Gota*) Separator

Development of an efficient separator for gota is required. As repeated passes of grains are given for complete removal of husk, separation of gota generated in a particular pass will reduce powdering of grains in subsequent processes. Screens are sometimes used for the separation of gota on size-basis. However, due to a very small size-difference, the screen graders are not found to be effective. Laboratory experiments have revealed that if gota is separated after each rolling operation, it results in increased *dal* recovery of 3 to 3.5 per cent over the prevailing milling techniques.

Future Research Needs

Breeding of Varieties with Desirable Milling Characteristics and Other Requirements

Development of varieties of desirable milling characteristics while breeding for improved varieties is essential as the desired impact of increased agricultural production will not be felt if it is lost at the processing stage. The pulses of different varieties may give different milling yields due to the varietal difference. Hence every effort is needed to study their properties and develop postharvest processing strategies. The desirable characteristics in the pulse varieties include uniformity of shape and size (preferably round), thinner husk, less force required to separate husk from pulse, loose binding between cotyledon for easier separation, resistance to storage insects and pests etc. The pulse should be screened for its milling and storage quality. The studies on engineering properties, drying and dehydration characteristics, pre-milling treatments and process development need to be taken up in an appropriate manner.

Prophylactic Treatment

Prophylactic treatments to prevent insect and microbial infestation should be a part of the strategy. It is evident that these treatments like conditioning and drying, light roasting and heating enhance the milling quality. However, optimization is essential in each case and the treatments need to be verified on large scale for different varieties of pulses before giving recommendations.

Substitution of Edible Oil by Other Suitable Products

It is an area in which can be made to try some other alternatives like application of low concentration of chemical solution ($NaHCO_3$) or vinegar (fermented sugarcane molasses) or some non-conventional oil application to eliminate the use of edible oils in the pretreatment process.

Efficient Conditioning Process

The grains are conditioned by various methods for loosening of husk, ease of milling and getting higher *dal* yield. The method includes hydration, drying, scouring, oil treatment, roasting, chemical treatment, tempering etc. Inactivation of gummy layer which exists between the seed-coat and endosperm is aimed during conditioning for loosening of husk. However, the last and most economical proposition is yet to be arrived at. The comparative study of different methods used for conditioning may be done and the most useful method may be recommended/extended to the users after standardization. The standardization for sequence of operations for each pulse crop should be carried out as it would save the losses to a great extent.

Composite Pulse Milling Equipment

Like the rice milling equipment, there is a need to develop composite pulse milling equipment which should undertake different unit operations without recourse to excessive handling. The adoption of existing grain milling equipment with minor modifications for pulse milling would be useful to reduce the investment and cost of milling.

Pre-cooked Pulses (Instant Pulses)

Pulses are cooked after soaking to improve the availability of nutrients as a result of destruction of certain anti-nutritional enzyme inhibitors and flatulent factors. But excessive cooking leads to lowering of digestibility of proteins. The protein efficiency ratio (PER) of common beans is improved as a result of cooking. Pressure cooking or steaming has been found to be better to achieve destruction of most of the anti-nutritional factors. Prolonged cooking lowers the

PER of legume proteins and also certain vitamins. Hence optimization of cooking to achieve proper treatment is very important. Commercialization of instant-cooking *dal*-mill gives relief to employed ladies and also generates employment by opening of cottage and small scale industries. The instant *dal* technologies so far developed should be given a careful consideration. The *dal* analogues from broken and powder-by products of pulse milling could be prepared by using extrusion-cooking technology.

It can thus be construed that:

1. Pulses form an integral part of Indian dietary as these are rich sources of protein. Their major importance is as supplements to cereal based diets. These not only add to the quantity of protein in the diet but also improve its quality by balancing the essential amino-acid pattern in the mixed diets. India is a major producer of pulses in the world. A number of varieties of pulses are grown and consumed in a variety of food products after suitable processing. Apart from common pulses like bengal gram, red gram, black gram, and green gram, a good number of other pulses which are generally referred to as minor pulses like cowpea, horse gram, lentil, moth bean etc. are also produced and consumed in different regions of the country. The present annual production of pulses is about 14 million tones and this has remained stagnant since the last two decades. As a result the availability of pulses has fallen to about 35 g per head per day. This is about half of what is actually required for proper nutrition.
2. Interaction of grain and machine parameters in relation to yield and quality of milled products is another area which requires attention. Milling systems of future may involve heat treatment as a promising approach. Heat treatment is invariably involved in pulses drying. Therefore influence of temperature and time combinations at various grain moisture on pulse nutrients, pre-treatments and milling quality need to be systematically investigated. Various pre-treatments and other processing efforts may involve losses of nutrients which need to be estimated.
3. Under every milling process certain mixture of unhusked and fully husked grains are expected. The fully husked

grains must be separated to avoid their scouring in recycling. As of now, efficient separation techniques are not available. There is need of a proper conditioning system to enable dehusking and splitting without causing processing losses. Various conditioning systems under different names were developed and adopted by the pulses processing industries since 1930s. Like paddy milling, milling quality of a pulse cultivar developed is an important criterion, which contributes towards its acceptability for cultivation. Evaluation of milling quality requires standardisation of the technique and apparatus. This should be taken up as an important activity.

4. Improved system of conditioning, incorporating there-in all the good features of the existing conditioning systems came into being as a result of the research and development work done by various institutions and the industry. The advanced systems as available now eliminate the processing losses to a great extent. The quality of the product namely, the dehusked splits, is much better. No matter how efficient the process is, the broken and powder formation is bound to occur.

References

Krishnamurthy, K., Girish, G. K, Ramasivan, T., Bose, S.K., Singh, Karan and Tomar, R. P. S. 1972. A new process of removal of husk of red gram using sirka, Bulletin of Grain Technology, 10: 181.

Kurien, P.P. and Parpia, H. A. B. 1968 Pulse milling in India I- Processing and milling of tur, arhar (*Cajanus cajan linn.*), CFTRI, Mysore, 11 June.

Matanhelia, K. G.,.1980. Commercial *dal* milling in India. Proceedings of the International Workshop on Pigeon pea, Vol.-I. ICRISAT, Patanchery 15-19 December.

Mayande, V. M.,.1987 Pigeonpea (*Cajanus cajan* L.) milling Technology: Biochemical and Engineering Studies, unpublished M. Tech. Thesis, G. B. Pant University of Agric. and Technology, Pantnagar.

Reddy, V. P. R. 1981. Dehusking of arhar grain (Cajanus cajan). Unpublished M. Tech. thesis, G. B. Pant University of Agric. and Technology, Pantnagar.

Saxena, R. P., Singh, B. P. N.; Singh, A. K. and Singh, J. K. 1981 Effect of chemical treatment on husk removal of arhar (*Cajanus Cajan*) grain. Paper presented at the Annual Convention of the Indian Society of Agricultural Engineering (ISAE), New Delhi, India.

Srivastava, V.; Mishra, D.P.; Laxmi Chand, Gupta, R.K. and Singh, B.P.N. 1988. Influence of soaking on various biochemical changes and dehusking efficiency in pigeon pea (*Cajanus cajan L.,)* seeds. Journal of Food Science and Technology 25,: 267-271.

Verma, P.; Saxena, R. P.; Sarkar, B. C. and More, P. K. 1993. Enzymatic pre-treatment of pigeon pea (Caajanus cajan) grain and its interaction with milling. Journal of Food Science and Technology. 30: 368-370.

Chapter 6

Enhancing Food and Nutritional Security through Postharvest Management and Value Addition in the Present Era of Globalization

S.P.S. Guleria

Department of Postharvest Technology,
Y.S.P. University of Horticulture and Forestry,
Nauni, Solan – 173 230 (H.P.)

The term globalization is a multi-dimensional concept and it incorporates internationalization of various aspects of economic life like trade and industrial relations, technology growth and transfer of technology, exchange of ideology and culture, investment by transnational companies and movement of capital and human resources. The effect of these different forms of globalization has been to create patterns of employment, expectation and demand for technology, skilled labour as well as commodities which can hardly be met solely through national effort and polices. The developing countries are now-a-days exposed to tremendous possibilities of growth and development through easy access to global link and through imbibing the virtues of the developed industrialized countries. The broadening of world market and free movement of capital and technology have greatly lifted the barriers to development to which the poorer countries were previously subjected to in the form of restrictions on trade, limited domestic market and inadequate capital resources to boost up the production system.

The world is rapidly changing with the globalization and liberalization of economy, rise of new conflicts, failures of economic growth in sub-saharan Africa, Latin America, Caribbean and Arab states and transition to free market economic system. All these changes put added stress on the lives of the people who already suffer deprivation in many aspects of their lives. Pointing out the inadequacies of our economic reforms, Amartya Sen laments that India had suffered a great deal in the last 50 years by ignoring the need for integration of social sector development. The tendency to neglect social development, in particular has very far reaching consequences (Singh and Gaur, 1999). The willful neglect and exploitation of the social sector will further worsen the condition of the poor and starvation deaths would continue to rise in many parts of the country.

How do we ensure that all people get access to good quality, safe and nutritious food? Swaminathan (2001) stated that sustainable food security needs concurrent attention of three components viz; i) availability in the market, which is a function of production; ii) access to food, which is related to purchasing power and hence to jobs, or livelihood and iii) absorption of food in the body, which is determined by access to safe drinking water, environmental hygiene and primary health care. Guleria (2003) presented an over-view of food security in the 10th All India People's Science Congress, Shimla, highlighting it in the global and Indian context and ensuring strategies for the future, the details of which are mention below:

Food Security: Global Context

"Food insecurity is a complex issue having several dimensions such as poverty, unemployment, famine, gender discrimination, starvation, food and nutritional practices, human growth, political elements, natural calamities and so on" (Ghosh, 2000). Poverty has been recognized globally both as a cause and consequence of food insecurity. The household is also not a homogenous unit, with women and girl children tending to suffer more from endemic hunger. Besides, economic inequity is widening nationally and globally. UNDP's Human Development Report (1999) points out that OECD countries with 19 per cent of the global population have

71 per cent of global trade in goods and service, 58 per cent of foreign direct investment and 91 per cent of all internal users. The world's 200 richest people more than doubled their net worth between 1994 and 1998 to more than one trillion dollars. At the same time, 1.3 billion people–two thirds of whom are women, live without access to balanced diet, clean water, sanitation, primary health care and basic education.

The Universal Declaration of Human Rights (1948) suggested that everyone has the right to a standard of living adequate for health and well being of himself and his family including food. The necessity of having food security at the global level was first recognized at the 1974 World Food Conference, and in spite of the initiatives taken by various agencies at the world level, poverty continues to be with us in huge proportion. Six infants of every 100 do not see their first birthday and eight do not survive to their fifth. Of those who do reach school age, nine boys in 100 and 14 girls do not go to primary school (World Development Report, 2001). Now with adequate global food output, 800 million people are hungry and this will continue to grow unless action is taken now.

World over, an estimated 840 million people are chronically undernourished. They are unable to grow or obtain enough food to lead healthy and active lives. This also includes more than 200 million children under the age of five who go to bed hungry every night, lacking the essential calories and protein that their growing bodies require. Hunger not only cut short the lives of individuals, it also damages the peace and prosperity of the nation. As an off-shoot of chronic hunger, the nation faces enhanced conflicts and social unrest, often accompanied by blatant misuse of fragile natural resources. Among poor people, wherever they are, women and children, the elderly and the sick are most likely to suffer from malnutrition. Most of the poor families in the world spend almost 80 per cent of their income on food, leaving little for housing, health, education or the management of natural resources. The first cause of hunger and malnutrition is poverty. People fail to eat enough food because they can not afford to produce it or buy it.

Food Security: Indian Context

Food as a basic human right as well as the basic necessity for mankind has been recognized in India as is reflected in Article 47 of our constitution which enjoins upon the state "to raise the level of nutrition and standard of living of people and the improvement of public health". What causes hunger in India currently is the wide prevalence of poverty and the inability of a large section of the population to buy food (Sen 1999). Sen further points out that though India has achieved self sufficiency in the availability of food grains due to the remarkable success of the green revolution, while assessing this in the light of limited purchasing power of the Indian masses, it is very low. Around 50 per cent of the people in India are estimated to be starving for want of adequate purchasing capacity and this proportion of starving poor is rising. It is realized that in low income countries the vulnerable sections continue to be unprotected even if the country attains food self sufficiency.

During 1999-2000, food grain production touched 200 million tonne mark; together with a very significant achievement in horticulture sector with all time record harvest of onions of the order of 4.76 million tonne and India becoming the largest producer of fruits and second largest of vegetables in the world (Sarkar, 2001). To add further, India has emerged as the largest producer of milk with 78 million tonne production level before entering into 21st century. However, the population of our country has been expected to grow at the rate if 1.8 per cent and projected to reach 1350 million in 2011. To meet the demand of our growing population the country will need food grains, milk, vegetables, fruits, fish edible oils to the tune of about 260-264, 130-150, 137-193, 80-106, 10-14 and 12 million tonnes respectively, to provide adequate nutrition to 1.35 billion people by 2020.

Strategies to Ensure Food Security

In developing countries where majority of the population still depends upon agriculture for their livelihood, this sector needs to be improved. These countries have to mainly depend upon domestic production for meeting their demand. When the production possibilities are fully exploited, it will result in raising production, increasing employment opportunities, enhancing incomes and thereby increasing the purchasing power of the poor. Growth in food and agriculture has been the basis of economic growth, higher per capita income and better diet in most of the countries. For example, China's extra ordinary annual economic growth rate of 9.5 per cent during the 80s to 90s was preceded by rural and agricultural policy reforms of the late 1970s and early 1980s.

Indonesia and Thailand also experienced strong agricultural growth prior to the period of non-agricultural growth which continues today.

Advances in agricultural technology can have profound effect on the lives of poor people. Introduction of quality protein maize in Guizhou, poorest province in China, has changed the lives of people miraculously. Until recently, the annual income was less than $50 per capita and for upto three months a year, the families had virtually no food. Quality protein maize is higher yielding than conventional varieties, but more important, it has higher levels of two essential amino acids vital for the growth of children. Today, the local people are better fed and surplus maize has been used to produce pork, increasing food security and disposable income. Having transformed the lives of 25,000 families in Guizhou, cultivation of the hybrid variety of maize is being adapted in the neighbouring provinces. In Africa, adoption of improved maize raised yields to an estimated level of 12-14 per cent with gains as high as 40 per cent in areas with favourable conditions (Byerlee and Heisey, 1996).

Despite advancement, cereal yield in developing countries has been declining steadily from 2.9 per cent a year in 1967-82 to 1.8 per cent in 1982-84 with demand for food grains predicted to increase by 59 per cent in the next 25 years. The developing countries should undertake concerted efforts in capitalizing the advances in agriculture. Priority must be placed in increasing the momentum for advancement in agriculture to ensure food security. According to World Watch Institute Washington, average annual increase in food grain production on a global basis was 30 metric tonne, between 1950-54, which dropped down to 12 metric tonne during 1989-92 and it is projected to drop further down to 9 metric tonne between 2000-2050 when the world population would rise to an estimated 8.90 billion.

Boosting Agricultural Production

Presently the crop yields in India are quite low. The natural resources which have direct influence on food production are under gross neglect and under utilized. In the country, there is over 100 million hectare land classified as wasteland and about 65 per cent rain-water is wasted, over 80 per cent of the livestock are less productive and around 50 per cent forests are denuded, while over 55-60 per cent of rural people are under employed (Mathur 2000). In such a situation we should make every effort to exploit these resources fully so as to boost agricultural production.

A development policy that focuses on increasing production of goods without addressing sustainability if resources on which production is based will definitely run into declining productivity. To be profitable, agriculture must be guided by modern researches. Agricultural trade liberalization can also help to reduce poverty and food insecurity. It contributes to food security in a number of ways, by making up the difference between production and consumption needs, reducing supply variability featuring economic growth, making more efficient use of world resources; and by permitting global production to take place in those regions more economically suited to it. There is a also the possibility of increased production and a resultant increase in rural labour demand and rural real wages. This will help to reduce the food insecurity problem by increasing economic access to food. Further, reliance on trade to bridge the shortfall between production and consumption is not free from risks. The risks may be uncertainty of supplies and world market price instability. Access to global market in terms of food security requires access to the required foreign exchange. Given the past level of export performance and the projected global economic conditions, a sustained export performance for most developing countries including India, appears to be uncertain (Selvaraj and Jayaprakasham 2001). The challenges of producing more food from less land and water can be met by developing integrated farming system, which require agricultural intensification and value addition.

Generating Employment and Income

Over the past five decades, yield increase in the major food grains throughout the world has been substantial. Improving agricultural productivity tends to increase farm employment and provides an important stimulus to the overall economy. This will help to increase total household income and will ensure better nutrition. As there are very less opportunities of increasing cultivable land, there is no option except to increase productivity of agriculture through evolution of high yielding varieties. Evolution of short-duration varieties and suitable cropping pattern will boost agricultural production per unit area. Regular food supply and stable prices create food security. It will also be necessary to provide the basic necessities such as drinking water, education and health facilities and organized rural marketing to safeguard the rural poor.

Increasing Food Production and Supply at the National/ Household Level

There is vast scope of increasing agricultural production per unit not only from irrigated areas but also from rainfed and semi-arid region, improvement in storage/processing and preservation of food and its quality control. Transgenic crops could help to boost food production while promoting environmentally sound cultivation practices. Transgenic crops can also be tailored to marginal agro-ecological conditions. Rising purchasing power in rural areas leads to rising demand for both agricultural and non-agricultural products (Quaim, 2001).

Reducing Incidence of Poverty

The major problem facing the rural sector of the developing countries is the persistence of poverty and its associated problems of low productivity, under-employment, rural-urban income disparity, un-organized marketing, inadequate infrastructure, poor water, health and educational facilities, poor living conditions etc.

Countries that have done well in agriculture (such as Indonesia, Malaysia, Republic of Korea and Thailand) have experienced radical decline in rural poverty and improved food security. These countries have followed broad based economic development with high levels of investment in agriculture, education and infrastructure and have succeeded in increasing food security and drastically reducing poverty. So, what is required in the context of developing countries is the policy reforms on the part of the government to expand investment in rural infrastructure, education, health, agricultural research and improved food production technologies.

Expansion of Human Resource Capabilities

The progress of a country depends largely on the progress of the people. In fact, poverty alleviation is not possible without overall development. Co-operation is one of the mechanisms by which human resource can be developed/organized and material resources mobilized. Cooperative societies are to be formed in all villages and the members are to be imparted training in the areas of farm technology, plant protection, harvest and postharvest operations.

Food Storage

One third of the fruits and vegetables produced in India is wasted for lack of adequate storage facilities which would help in greater availability of food for distribution. If food storage-based coping mechanism is adopted, storing the staples in peak season and selling them during economic crisis will benefit the poor. In India, food storage facilities are limited and a lot of food is wasted or eaten by rodents, insects, pests and other parasites. Thus suitable arrangements are to be needed to reduce storage losses.

Public Distribution System

Although there is a vast network of public distribution system arrangement by the government, the procurement, buffer-stocking and storage is not efficient and there is a general complaint of low grade agricultural commodities and irregular distribution by these shops. Government should not only make such shops available in rural areas but also ensure the quality of grain and its availability to the consumers throughout the year. Recent studies on PDS reveal that there has been a decline in the share of income spent on cereals among households in lower expenditure categories. So, cereals of good quality at cheaper rates should be distributed to enhance the food and nutritional security.

Strengthening of Employment Programmes

The central/state governments have initiated a range of poverty alleviation programmes with a view to empowering the rural poor to enable them to participate in agricultural growth. These include both wage and self-employment programmes. Field studies have indicated that only a marginal segment of the sample households were able to find work under the JRY. Further, because of widespread illiteracy, there is no awareness among people about government programmes. The condition of women labour and farmers has to be strengthened because they contribute a lot to the well-being of the household.

New Dimension in Fruit and Vegetable Processing: Challenges and Innovations

The importance of fruits and vegetables in human nutrition is very well known. These are rich sources of vitamins, minerals, salts, low acids, bases, reducing and non reducing sugars, alkaloids, fibers, pectin, fats, enzymes, etc. which help to regulate body processes and also give protection us from the ill effects of many diseases, disorders and malnutrition. The daily requirements of the some of the essential nutrients can be very well met, if a person consumes about 85 g fruits, 75 to 125 g leafy vegetables, 85 g other vegetables and 85 g roots and tubers everyday on a balanced diet pattern.

Vegetables are quite costly these days and quite often their purchase becomes prohibitive for a poor. At the same time, 35 to 40 per cent of our produce still goes waste at production sites, collection centers and mandis due to physiological, physical, biochemical, microbiological, and technological factors. The value of fruits and vegetables must be realized particularly by women and lactating mothers, so that they can save their growing children from the diseases caused due to malnutrition.

About 700 million people all over the world go to bed hungry and 800 million suffer from chronic malnutrition. Almost one-third of the world's chronically mal-nutritioned people live in India. About 200 million children under the age group of five years suffer from malnutrition. An alarming rate of population growth, poverty, ignorance of people about balanced diet, consumption of unhygienic food, postharvest losses, etc. are the main factors. The magnitude of malnutrition is discernible in the hilly regions of India.

The Challenges

In my opinion, product promotion technologies and value addition exercises which may be designed suiting to the needs and food habits of ruralities can transform the rural economy and can save them from malnutrition and its related diseases. It is high time to establish close linkages with vegetable growers and gear them up for adoption of preservation culture, processing culture and solar drying culture so as to bring them home to adopt the old art of making *achars*, *mulikanda*, *murabbas*, *juices*, *chutneys*, concerntrates, *pethas*, sprouted pulse grains, *madirus*, *koftas*, *manees*, *khattas*, *meethas*, nuggets, *pappars*, *chats*, stuffed foods, deep-fried foods, *sarson-ku-sag*, *berrarair-roti*, *missi roti*, *bethein*, *patore*, hot ash baked potatoes, onions, and *arvi*, starchy foods baked in wood ash, cooking fruit or vegetables in *lussi*, curd, or *matha*, making *chopps*, *tikkies* or *pakoras*, or *bhujus*, dried or cooked foods from green leafy vegetables, making *kheer*, from cabbage, cauliflower, carrots, *patha, shakarkandi;* making *halwas* from carrots, potatoes and starchy tubers; stuffing fruits vegetables with *masalas, ambchur*, and *anardana* powder; making *barfis* and *ladoos* from *tindas,* long gourds, round gourds and *pethas;* drinks from melons; *sund* prepared from ginger, nutfruits, sugar and *khoya chambachukh*, chilli powders, green chilli paste with *pudina* and *hara dhania* etc. which are age old techniques with our household hill masses and which can be easily improved upon. The scientific basis behind the preparation of these existing traditional or ancient foods may be disseminated to the ruralities so that the indigenous technological knowledge (ITK) may be kept alive by the ruralites. For instance, the ruralites throughout Himachal Pradesh use methi seeds while preparing butter fat merely because it imparts flavour to it only. However the scientific basis behind the adoption of methi seeds is the presence of tocopherol or vitamin E in these seeds which being soluble in fat and fatty substances acts as an anti-oxidant to prevent the *ghee* from going rancid. Moreover, desi-ghee used to be fortified with vitamin-E as it is anti-sterility in nature. Thus the ruralites need to be educated about the scientific basis behind the use of *methi*-seeds so that they may also use these for the preservation of all fatty foods such as oils from *sarson, taramira, til, alsi* etc. so as to prevent all fatty foods from going rancid and also fortifying these preparations with vitamin-E. The revival of old preservation culture will add longevity to the vegetables, which have a very short period of availability and shall also avoid wastage at production sites.

Innovations

Due to the recent influx of foreign food culture like *Pizzas, Noodles, Salamis, Hot Dogs, Burgers* and related fast food items by the media in our country, there is a direct attack on the traditional food culture like *Makki-ki-roti, Sarson-ka-sag, hari mirch pudina* sauce, *chachh-matha*-vegetable preparation, *madiras* full of *petha* and nuts, stuffed *tindas*, stuffed capsicum and stuffed bitter gourd, *okra* and *pakoras* of different kinds made by the incorporation of fresh vegetables and related delicacies. There is likelihood that these house hold treasures of Indian kitchen may become extinct as the same media is not making any effort to popularize and promote these transitional food cultures. Thus, these foods may be popularized along with the new foods or innovative foods.

The innovative foods or new foods can change the developmental mode of the present market expansion scheme or expansionist thinking process of the media. Any new introduction into the market can bring a behavioral change amongst the consumers. Once the consumers get accustomed to the taste, desirable characteristics and utilization mode along with penny saving statistics involved in such new innovations as well as the nutritive value of the new market introductions or new food preparations, the market demand will automatically get established.

Innovative Foods Developed

Some of the innovative foods developed by Guleria, (2004) have been described below:

Tomato-potato Nuggets

Blanched tomatoes may be rubbed on a stainless steel mesh or sieve to get the maximum juice and pulp devoid of peels, hardcore and seeds. This pulp is concentrated to one-fourth of its original volume under gentle flame (simmering temperature) or around 95 degree celsius. Thereafter, the potatoes are boiled, peeled and grated. The grated potatoes are made into a fine quality dough by knidding gently so that the knidded mass does not contain any grittiness. This fine dough of potato is mixed with concentrated tomato pulp in the ratio of 1:1. *Haldi, mirch, masala* etc. may be added as per one's taste, likes and dislikes, suitability, economic viability or pocket affordability. Then by using *hing*, the nuggets may be made, sun-dried and packed in cloth or polythene bags or may be stored in earthen pots.

These tomato-potato nuggets may be used when tomatoes are not available or during the off-season, when the price of the tomatoes sores very high. These nuggets when fried get disintegrated into the prepared *masalas*, and the tomato portion of the nuggets imparts sour taste and colour to the food preparation; whereas the potato portion of the nuggets renders thickness to the food slurry. In the similar manner, the nuggegts may also be prepared from the tomato pomace containing peels, hardcore and seeds etc. which may be grounded in a warring blender or grinder before mixing with potato dough etc. The utilization procedure of tomato pomace–potato nuggets is the same as narrated for tomato-potato nuggets.

Tomato Leather and Tomato-powder

Tomato concentrate may be spread over an oil-smeared plate, aluminum sheet, slate, wooden tray, *khajoor*/bamboo mat or polyethylene sheet and then sun-dried. The spreading procedure may be carried out in layers so as to get a brick like structure. It is called tomato leather. It can also be prepared by using *masalas, hing* etc. The leather making can also be done in an oven at 45 to 50 degree celsius by treating it for half an hour so as to keep the moisture content around 60–65 per cent. By the addition of wheat starch (Nashasta) or poppy seed powder (Sieul atta) to the tomato concentrate in the ratio of 2:1 followed by drying, the tomato-powder can be conveniently prepared and it may be preserved by the addition of *hing* powder or potassium metabisulphate which may be incorporated in the powder @one gram per 100 grams of tomato powder. This tomato powder can also be made by the use of *maida* and maize flour. The maida or maize flour may be strained through a fine meshed cloth at least 2 to 3 times in order to get a fine flour before incorporation in the tomato concentrate. It may be utilized for making soups and curries.

The tomato leather may be submerged in lukewarm water to dissolve it to make pulpy or juicy mass for utilization in food preparations or for making ready-to-serve hot juice or soup. The tomato powder may be hot-cooked in appropriate quantity of water to make a soup of desirable consistency which may be served along with butter as a hot delicacy. The tomato leather may be conveniently stored in polyethylene bags, dry earthen pots or in cloth bags at a dry place. The tomato powder may also be packed in polythene, cellophane or *mitti-ka handoo* duly sealed with a lid and stored in a dry place. Silica-gel wrapped in a cloth bag may be kept in packed powdered-food so as to save it from gaining atmospheric moisture or storage odours.

Gourd Slurry as the Base Material for Canning of Mushrooms

The peeled long gourd or round gourd may be grated and then cooked at simmering temperature (95 degree celsius) for about half an hour. Thereafter the heat-treated mass is strained. The filtrate thus obtained is concentrated to one-third of its original volume with the addition of one percent rock salt. The salted concentrated mass is the desired slurry in which steam blanched mushrooms may be canned. The advantage of this slurry usage is that when the mushrooms are drained out, the slurry serves as a stock solution to make curry for the mushroom food preparation. Generally the mushrooms are canned in a saline solution and during the mushroom food preparation, the saline solution is thrown away. But in the present innovation the slurry used as a base material for the canning of mushrooms is a food item in itself and is used in mushroom food preparation.

Gourd Nuggets from the Pomace Left Over after Slurry Making Process

The long gourd or round gourd pomace left after straining the juice may be exploited for making nuggets using the desired quantity of *haldi, masalas*, black pepper powder, chilli powder, *hing* etc. The nuggets may be sun-dried or dehydrated at 45-50 degree celsius for half an hour at 60 per cent relative humidity. Else the grated pomace may be used for making *koftas*. These nuggets may be used in food preparations when added to fried curd or to *karri* made from *lassi* or thick fried tomato curry. These may be stored in a similar way as tomato-potato nuggets.

Jelly from Watermelons

The pinkish edible portion of the watermelon alongwith juice is grated and after separation of seeds is put in a warring blender or a mixer to get a uniformly broken mass. Thereafter equal quantity of sugar is added to it and this combination is heated at a simmering temperature to bring it to a 40 degree brix consistency. After that half-kilogram more sugar is added and heating at around 95 degree celsius is continued. When the gel-like consistency (65 to 68 degree brix) is obtained, 2 grams of citric acid is added. Thereafter, after slight heating the flame is put off. The jelly is ready which may be filled in glass jars as a hot gel-mass and allowed to cool. When cooled to room temperature, the jar is lidded. Instead of glass jars, china clay jars may also be employed for filling. These jars may be stored at a cool and dry place or cold stored in a refrigerator. Before lidding, paraffin wax may be poured on the mouth of the jar to prevent interaction of out side air with the jelly contents. If the jelly is weeping, then one gram of pectin grounded with four grams of sugar may be added to the hot gel-mass being cooked at 95 degree celsius to get a well-set jelly. This jelly may be eaten along with *paranthas* or *chapaties*, *makki ki roti*, *bajra* or *rice-roti* and can also be applied on bread/toast. This jelly may also be utilized for dressing of puddings, cakes, ice creams, custards, *halwas* and sweets.

Jams from Musk Melons

The jam from muskmelon pulp is made and stored in the similar manner as the jelly from water melons as explained earlier.

Watermelons Musk Melons Admixtured Jam Using Sugar

The watermelon and the muskmelon pulp may be mixed together in the ratio of 1:1. This admixture may be beaten in a warring blender or a mixer to get a homogenous mass. Two parts of sugar is added to one part of pulp. This is then heated at simmering temperature (95 degree celsius) till thick gel-like mass is obtained or when the consistency of the gel reaches 65–68 degree brix. At this point of juncture, the citric acid (1 to 2 grams) may be added and after slight heating the flame may be put off. This jam which may be filled in glass or china clay jars and allowed to cool to room temperature. Hot paraffin wax may be poured at the mouth of the filled jar, which may then be lidded and kept in cool dry place for storage or stored in a refrigerator. The consumption procedure for the jam is similar to the one narrated for the jelly.

Diabetic Jam Using Melon and *Bael* Pulp

For making the diabetic jam devoid of sugar, the watermelon pulp (one part) and muskmelon pulp (two parts) may be beaten in a warring blender to get homogeneous mass. To this admixture add *bael* pulp in the ratio of 2:1 and heat the melon–*bael* admixture at 95 degree celsius to get a gel-consistency (65 to 68 degree brix) with the addition of 2 grams of citric acid at the gel point. This jam may be filled in glass jars or china clay jars in the similar manner as narrated above for the sugar containing jam. It may be paraffin waxed and stored in a cool dry place or in a refrigerator as narrated for the jelly/jam.

Hot Wood-ash Baked *Arvi* Slice Pickled in *Rye* and *Galgal* Juice

The *arvi* may be kept covered in hot wood-ash for about one or two hours so as to bake it to attain edible characteristics. The peel may be rubbed off and then sliced. The *arvi* slices are then fried in mustard-oil in a non-sticker pan or *tawa*. To one kilogram of fried slices 200 g of rock salt, 2 to 3 g of powdered *rye*, 500 g of *galgal* juice and 50 g of jaggery are added and thoroughly mixed. The entire admixture is then put in a glass jar or china clay jar the mouth of which is covered with a muslin cloth. This pot is then treated in the sun daily for about a week after which the pickle gets ready for eating. The muslin cloth is replaced by the lid of the pot. The pickle may be stored at a cool and dry place.

Bitter Gourd Peel as Colouring Agent for *Burfis, Chats* and *Chutneys*

Before stuffing, the bitter gourd is generally peeled off and the peels are discarded. It has been found that the peel of bitter-gourd is a very good edible and colouring agent for *burfis, chats,* and *chutneys*. This peel may be grounded in a pestle and mortar or warring blender or a mixer to get finely grounded homogenous uniform peel mass. When the *burfi* has been prepared, this peel may be added as a coloring agent to get an organic *burfi* devoid of synthetic colours. This natural colouring agent is useful to those patients who suffer from allergies due to the synthetic dyes and colours used in the sweets. For use in chats and chutneys, the bitter gourd peel may be grounded to a fine consistency along with fresh coriander leaves and fresh green chillies. This admixture gave a mouth watering flavour to the *chats* and *chutneys*. For further enhancement of taste, a pinch of mango powder or *anar-dana* powder and mint powder or mint fresh leaves' paste may be added during colouring paste preparation. The colouring agent paste may be sun-dried on a wooden plate or waxed sheet or polysheet or slate etc. to get a powder which may be packed in food grade polythene bags or cloth bags and may be stored in a cool and dry place.

Tuti-fruity from Petha and Melons

The *petha* or melons may be cut into half inch pieces having different shapes. These pieces are then kept in lime water overnight to get the surface of the pieces hardened and enriched with calcium of lime water. These pieces are then tied in a muslin cloth and blanched in hot boiling water for five minutes. After blanching, the pieces are spread over a white sheet or bamboo basket or bamboo net or mosquito net for one hour at room temperature to dry off the surface moisture. Then these blanched and dried pieces are cooked in sugar syrup (40 degree Brix) for half an hour at 95 degree celsius. The bitter-gourd dried peel or fresh peel paste may be tied in a muslin cloth which may be dipped in the sugar syrup containing *petha* or melon pieces so as to impart a natural green tinge to these. The pieces are then drained-off over a sieve. These pieces are called tuti-fruity which may be dried in the sun and filled in jars and stored at a cool and dry place. These may also be packed in stainless steel containers. The tuti-fruity is made into a candied product by spreading over it dry powdered sugar during the sun-drying process. The tuti-fruity is used as a dressing material for ice creams, cakes, *halwas, kheers*, custards, *pans* etc. and may also be utilized for making *meethas* during *dhams* and ceremonial food preparation, *yagyas* or festivals. The candied tuti-fruity is very good chewable toffee-like food to be served to the children.

Herbal Preservation of Vegetable Juices, Soups and Concentrates

The mint and basil extracts added @ ten to twenty grams per 250 to 500 ml of vegetable juice, soup or concentrate, serves as a preservative or and is similar in effect potassium meta-bisulphite or sodium benzoate. The mint or basil extract may be added to the vegetable juice, soup or concentrate at the end of prepared food before bottling or filling in jars. The mint and basil leaves are finely grounded into a paste and its uniformly grounded homogenous mass or its extract may be utilized for addition to or for incorporation in vegetable juices, soups or concentrates as a preserving agent in a freshly prepared state only.

Okra Extract as Filtration Aid for Honey Clarification

Fresh okra may be cooked as a whole fruit at simmering temperature (95 degree celsius) and the extract sieved out on cooling. The extract when spread over a muslin cloth or when smeared over it makes a very good filtration agent for the clarification of honey. The honey containing jar is treated indirectly in a boiled water containing vessel so that the temperature of the treated honey does not rise above 40 degree Celsius so that the honey enzymes do not get destroyed which determine the medicinal properties of honey. This treatment is necessary to liquify the honey to increase its fluidity. The fluid honey is filtered through the muslin cloth smeared with okra extract. The filtered honey does not get crystallized due to the colloidal phenomena. The nuclei of the dextrose present in the honey get separated into colloids and do not coagulate to form crystals. Crystallization is the major defect in honey which is due to the presence of bonded-water which gets incorporated in the honey network during the honey extraction procedure from the honey-combs. This bond-water is difficult to be removed otherwise. So, the organic fitration agent like the bhindi exctract has the capacity to remove the bond-water due to colloidal phenomena. The bhindi extract may also be applied on the filters of the filter press as filtration–aid which helps to form honey of International Grade. This fitration aid process has been commercially employed by M/S Him Herbs, Raja-Ka-Bag, Nurpur, Distt. Kangra, Himachal Pradesh (INDIA).

Combating Malnutrition and the Role of People's Science Movement

Vegetables differ widely in composition and nutritive value, depending on the part of the plant used as food. These are the cheapest protective foods one can buy. Vegetables are the rich sources of various vitamins and minerals. Green leafy vegetables like amaranth, spinach, methi, mustrard are very rich in calcium, iron and many of the vitamins particularly vitamin A, Thiamin, Niacin, Folic acid, vitamin C, as well as the dietary fibre. Vitamin A helps us in maintaining a normal vision whereas C and B group vitamins not only protect us from various infections but also keep our skin healthy.

Certain roots such as carrots and radish are not only rich sources of various minerals and vitamins such as iron, calcium, potassium, sodium, B-group vitamins and Vitamin-C; but also contain a good amount of dietary fibre which helps in preventing constipation. Barring starchy roots and tubers, other vegetables such as egg plant, pea, beans, cauliflower, drumstick, etc. provide a variety of vitamins and minerals in our diet. Drumsticks, for instance, are especially rich in calcium and vitamin-C. The edible peels and seeds of these vegetables are rich sources of dietary fibre. Further, these vegetables add colour and variety to our meals.

Potato, yam, tapioca, sweet potato and elephant food which belong to starchy food group are the rich sources of carbohydrates and are hence energy giving foods. These can in fact replace a part of cereals in our food provided they are utilized along with skin in the form of baked food. Apart from carbohydrates, these are also the richest sources of calcium and vitamin-C.

Hence, vegetables not only help us in maintaining a healthy skin and good vision but also provide protection against growth failure, loss of weight, protein energy malnutrition, weakness, lethargy, constipation, deficiency of fat soluble vitamins, rickets, osteomalacia, anemia, low Hb, oxygen carrying capacity, breathlessness, night blindness, kaeratomalacia, poor resistance against infections, *beri-beri,* a riboflavin sis, soreness of tongue, cracked lips, pellagra, diarrhoea, dermatitis, mental disorders, megaloblastic anemia, pernicious anemia, scurvy, soreness and redness of tongue, obesity, cancer, diabetes, dehydration and coronary heart diseases.

Vegetables can be incorporated in our diet in variety of permutations and combinations. Thse can be consumed as such *i.e.* in the raw form as salads, fruit chats, or as stuffings in *paranthas, kofta* curries or even along with pulses, curds etc. Soups and juices prepared from carrots, tomatoes, peas, spinach, coriander leaves, mint etc. not only add taste to the meals but are also good for infants, lactating mothers, growing children and invalids. It is, therefore, high time to realize techniques of preparing *kheers, halwas, burfis, ladoos* pickles, *chutneys, murabbas* and candies from vegetables and start a People's Science Movement to get the vegetables and their products included in our daily diets to protect the people from the menace of disease. The edible part of vegetables is often juicy, pulpy, sweet or sourish-sweet in taste. So, let us all make all efforts to convert these desirable characteristics of vegetables into an "income generation activity" by preparing new foods to flood the market to raise the economy of the farming community and the country.

Laying special emphasis on Himachal Pradesh, the author appeals to the concerned extension workers in different state departments of Agriculture/Horticulture/Rural development/ Science and technology/Social welfare etc., including the Himachal Production Council, HPMC, Women organizations and the NGO's to spread the message of "preservation Culture" among the rural women to increase purchasing power of hill masses and decrease malnutrition by increasing self consumption of vegetables and their products. These extension agencies are also desired to organize the rural or farmer groups into Self Help Groups, Youth Forums, *Mahilla Mandals* etc. to convert the value-added technology into income generation or marketing and trading activities rather than keeping the information confined to household activities only.

References

Byerlee and Heisey. 1996. World Development Report (2000–2001): 3.

Ghosh N.G. 2000. Food insecurity-the greatest challenge of the millennium. Indian Farming, Vol.50 (7):7-9.

Guleria, SPS. 2003. Food Security: Concept and need for Postharvest Technology efforts. Xth All India People's Science Congress, Shimla, Samarika (10-14 October 2003):203-213.

Guleria, SPS. 2004. Postharvest Technology for Vegetables. Published by Principal Farmers Training Centre, Sunder Nagar, Himachal Pradesh (January 1-21,2001): 1-15.

Guleria SPS. 2004. Fight malnutrition through value addition. Advances in Horticulture, Deep and Deep Publication Pvt Ltd, New Delhi, pp 291-303.

Mathur P.B. 2000. World Food Day-a millennium free from hunger. Indian Farming, Vol.50 (7): 16-18.

Sabri M.A. 2000. Poverty and Food Security Problems and Prospects. Kurukshetra Vol. 49 (3): 29-33.

Selvaraj V. and Jayaprakasham, T. 2001. WTO and Food Security- A Reality or Illusion. Southern Economist. (August 2001): 7-10

Sen Amartya. 1999. The Nobel Laurate Discusses Key Issues. Outlook (Nov. 15 1999): 29-32.

Singh J.L. and Gaur, K.D. 1999. Need for New Strategies to Eradicate Poverty, Manak Publications: 216-217.

Swaminathan, M.S. 2001. Science and Sustainable Food Security. Indian Farming Vol.50 (10). 4-7.

Quaim, M. 2001. Transgenic Crops and Developing Countries. Economic and Political Weekly Vol.XXXVI (32): 3064-3065.

Chapter 7

Drying Technology

D.K. Gupta

Department of Process and Food Engineering,
G.B. Pant University of Agriculture & Technology,
Pantnagar – 263 145 (Uttaranchal)

Drying is a very powerful unit operation for food preservation and development of value-added new products. The microbial deterioration is considerably reduced at low water activity. The quality of the stored product depends upon conditions of storage particularly, temperature and moisture content of stored material and gas composition of storage system.

Higher the moisture content and temperature, greater is the deterioration of the stored product. If the temperature becomes too high, the deterioration may again reduce because many harmful micro-organisms may not be able to grow at elevated temperatures. However, high temperature is not recommended for storage because higher temperature leads to other problems, especially, quality as the quality of the stored product is poor at high temperatures. Because drying lowers the product moisture content, it helps to extend the shelf-life of the product. The drying is also used to develop several improved stable ready-to-consume products such as dehydrated peas, dehydrated mushrooms, dehydrated grapes, onion powder, potato flour, dried meats, fish and eggs, milk powder, instant tea and coffee etc. to name a few. In many cases, the crop is harvested at its maturity level at a higher moisture content such as paddy and subsequently, its moisture content is reduced to safe storage level by drying.

Drying is a process of simultaneous heat and mass transfer and the conditions of the drying medium are important in the drying process. If any two psychrometric parameters of the drying medium are known, the others can be calculated using psychrometric chart or model. The main consideration in the drying process is to utilize the maximum heat energy for removing the moisture and the rise in temperature of the product should be minimum so that the process can be used for heat- sensitive products without seriously affecting their quality.

Advantages of Drying

Some of the advantages of drying are as follows:

1. The volume and mass of the product is reduced making it easier to transport and store at a lower cost.
2. The moisture content of the product is lowered thereby the water activity is reduced resulting in longer shelf-life.
3. For some products, the taste is improved.
4. Solid content of the product increases.
5. New products can be developed.

Drying Processes

Drying can be undertaken through the following three processes:

Single Kernel Drying

In this process, the entire kernel is exposed to drying medium. The rates of heat and mass transfer are higher.

Thin Layer Drying

In this process, small quantity of the product is dried. This is mostly practiced in laboratory work. For deciding the thickness that can be considered as thin layer, ASAE or other appropriate standards should be followed.

Deep Bed Drying

In this case, a bed of the product is dried in sufficient quantity. The deep bed may be assumed to consist of a number of thin layers one placed over the other. The fresh heated air comes in contact with the bottom-most layer first and output of the bottom-most layer becomes input to the layer above it and so on. The rate of drying decreases as the thickness of bed increases.

A number of simulation models for predicting drying rates have been developed over the years, in both the thin-layer and deep bed drying. These models are either theoretical or semi- empirical or empirical in nature.

Ways of Drying

Drying can be accomplished in the following ways:

Solar/Sun Drying

The energy for drying comes from the sun. The drying-air can be heated with the help of solar-energy or the solar-energy can fall directly over the product to be dried. This technique is weather dependent and can be used only when the sun is shining unless there is provision for storage of solar energy.

Storage Bin Drying

The grain is put in the bin and air at slow rate is passed through the grain. A plenum chamber (perforated floor) is provided in this type drying so that heated air comes through the perforations and passes through the grain. The drying is slow and may take considerable time to achieve the desired moisture. It should be ensured that this is practiced in such a way that the stored material is not spoiled.

Heated Air Drying

The air for drying is heated with a suitable heat source such as fuel oil or electricity or biomass or solar energy or waste materials. This heated air is passed through the food material to be dried. The system requires a blower for blowing the air, connecting ducts, heater, plenum chamber etc.

Ambient Air Drying

The unheated air at ambient temperature is passed through the grain. The air is not pre-heated. The drying may be slow depending on the relative humidity of air.

Fluidized Bed Drying

The bed of the product is fluidized in such a way that the material is not carried away but at the same time each particle is fully exposed to air. The rate of drying is faster The required drying is achieved in 1-2 hours compared to 8-10 hours in heated air drying.

Rotary Drying

These dryers rotate at slow speed and are mainly used for free flowing and granular solids. These are slightly inclined with the horizontal.

Conveyor Drying

The drying is carried out by passing heated air through the perforated belts, the direction of air could be upward through the product and then the direction could be reversed to avoid spillage of the material. Moreover, the relatively dried product may be piled on the belt to a greater thickness.

Pneumatic Drying

The drying-air velocity depends upon the product and is high enough to convey the product while drying it.

Cabinet Drying

These are insulated cabinets. The desired temperature is maintained in the cabinet and the food to be dried is put in perforated trays or any other suitable container. This way, the desired drying is achieved in the cabinets.

Osmotic Drying

The product is dipped into a solution of salt or sugar such that the moisture comes out of the product but the salt/sugar is gained by the product. The osmotic drying alone is not enough as it removes only part of the moisture and is followed by other methods of drying.

Freeze Drying

In this drying, the product is frozen first and the drying is achieved by sublimation of ice. The pressure in the drying chamber is controlled. Though this type of drying is expensive, the product quality is better. This method is particularly applicable to heat-sensitive products.

Vacuum Drying

The drying is achieved under vacuum *i.e.* under lower pressure. As the pressure is reduced, drying can be carried out at lower temperature. This way, it can be used for heat- sensitive products.

Spray Drying

In this method, powder is produced from liquid foods. Heated air flows concurrently with the atomized feed and the moisture is removed. The feed should be concentrated because more the solid content more is the production rate.

Drum Drying

There could be single drum or two drums. Concentrated feed is splashed over the surface of the drum. The feed forms a thin layer on the drum. As the drum, having diameter and length nearly equal, rotates at a slow speed (1-10 rpm) the feed is dried. The dried material is removed with the help of doctor's blade(s).

Combination Drying

In this process, two or more methods of drying are employed to achieve the desired extent of drying.

Classification of Drying on the Basis of Rate of Drying

Constant Rate Drying

In constant rate drying the rate of drying is constant with respect to time. The drying is achieved between the initial and critical moisture content of the product, if the initial moisture content of the product is higher than the critical one. Critical moisture content is the moisture content at which the rate of drying changes from constant rate to falling rate. In other words, this is the lowest moisture content to which a product can be dried under constant rate or the highest moisture content from where the falling-rate period begins. In fact, the initial moisture content of a large number of products is less than the critical moisture content indicating the absence of constant rate drying. This is particularly true of durables such as cereals.

Falling Rate Drying

The rate of drying decreases with time in this case. Such drying is done in the range of critical moisture content to equilibrium moisture content that is the lowest moisture content to which a given air can dry the product if exposed to the drying air for a long time. If a sample is exposed to given conditions of drying air, the product has a tendency to reach a constant moisture content depending upon temperature and relative humidity of the drying air and state of the process *i.e.*, adsorption or desorption. This is called equilibrium moisture content.

The rate of drying is governed by the rate at which the moisture within the product is diffused to the surface and is removed by the drying air. In falling rate period, the moisture within the product comes to the surface and is removed whereas in constant rate drying sufficient moisture is available on the surface of the product.

Theory of Drying

Drying is a simultaneous heat and mass transfer operation. In order to understand the drying process, knowledge of heat transfer, mass transfer, psychrometrics besides Mathematics is required. The equilibrium moisture content is the minimum moisture content to which a product can be dried with air of certain temperature and relative humidity. The equilibrium moisture content is different for different products. The drying of moisture takes place from the surface first and then the internal moisture travels to the surface and is removed. There may be another situation in the sense that the surface of the product may be partly wet.

Since drying is a simultaneous heat and mass transfer operation, the relevant set of partial differential equations needs to be solved. A number of investigators have solved the partial differential equations to describe moisture transfer during drying (Pabis and Henderson, 1962, Henderson and Henderson, 1968, Hussain, 1972). Analytical solutions have been suggested for steady state, equilibrium-type boundary conditions and constant values of diffusion constant (Pabis and Henderson, 1962).

Numerical solutions were also obtained by Chu and Hustrulid (!968) and Whitaker *et al.* (1969) for the diffusion constant being a function of temperature and moisture content. The moisture may move in the liquid form due to capillary forces or moisture concentration differences, or in vapor form due to moisture or temperature differences or in a combined form due to total pressure differences (Morey *et al.*, 1978). Since the diffusion models are complex and their solution is time consuming and the values of diffusion constants are unknown in many cases, several empirical and semi-empirical models have been developed for thin layer drying. One such model is called exponential model. A number of investigators (Simmonds *et al.*, 1953, Hustrulid and Flikke,1959, Henderson and Pabis,1961,Watson and Bhargava, 1974) have used this model which is based on Newton's second law of cooling and is described below:

$$MR=(m-me)/(mo-me)=\exp(-kt)$$

where,

MR= Moisture ratio

m = Moisture content (db) at time, t

me = Equilibrium moisture content, db

mo = Initial moisture content, db

k = Drying rate constant, per hr

t = Time of drying, hr

The exponential model was modified by Page (1949) by introducing another constant, n:

$$MR=\exp(-kt^n)$$

The values of constant k and n are temperature-dependent. If there is a systematic trend with the temperature, these constants can be related to temperature otherwise individual values need to be used at different temperatures. In general, it was observed that the constant k was related to temperature for a product by Arrhenius type equation (Kulshrestha, 1983).

Values of constant n were different for different products and temperatures. The average value of constant n is generally between 0.4 to 0.8 for most durables. Constant, n, was related to drying air temperature for soybeans by Overhults *et al.* (1973) as;

$$n = A + B\,T$$

For shelled corn, White *et al.* (1973) related n with dry bulb and dew point temperatures of air as:

$$n= A+B\,T + C\,T_{dp}$$

However, Misra and Brooker (1980) related n for shelled corn to relative humidity of air and initial moisture content by the following equation:

$$n = A \ln rh + B\,Mo$$

Flood *et al.* (1972) obtained a value of 0664 for n for corn and the following expression for k:

$$k = \exp(-xt^{y})$$

where,

$x = (6.0142+0.0001\,rh^{2})^{0.5}-0.01(1.8\,T+32)(3.353+0.001\,rh^{2})^{0.5}$

$y = 0.1245-0.0022\,rh+2.3*10^{-5}\,rh\,(1.8\,T+32)-5.8*10^{-5}\,(1.8\,T+32)$

rh = Relative humidity, decimal

T = Temperature, °C

Tdp = Dew point temperature,°C

Mo = Initial moisture content, db

A, B = Constants

Similarly, many researchers proposed different models for analyzing the experimental data. Many of these models are applicable in specific range of temperature and humidity. Some such models are given by Thompson *et al.* (1968), Chen and Johnson (1969), Kachru *et al.* (1970), Wang and Singh (1978), Sharaf-Eldeen *et al.* (1980), Sharma *et al.* (1982) etc. The trend of work in drying became such that the investigator would conduct experiments on the unexplored substance of his choice, collect experimental data and test the existing equations or models. If none of the existing equations could describe the experimental data, the investigator would propose a new model. This led to development of a large number of models.

Most of the thin layer drying models cannot be used without information on equilibrium moisture content (EMC) as the equilibrium moisture content is required for calculation of moisture ratio.

There are two basic methods of determination of equilibrium moisture content, (1) Static method and (2) Dynamic method. In static method, a controlled temperature space is maintained in ovens or incubators depending upon whether heating or cooling is required.

Ovens are used if the temperature to be maintained is above atmospheric temperature and incubators are used if the temperature to be maintained is below the atmospheric temperature. The controlled relative humidity conditions can be maintained by saturated salt solutions or acids because certain salts and the acids have the property that they maintain a fixed relative humidity at a given temperature.

The salt solutions are generally preferred over acids. Tables of such salts can be found in Spencer (1926) or any thesis/paper dealing with equilibrium moisture content. In the dynamic method, the product is subjected to the conditioned air. Although dynamic method is much faster, static method is more popular due to ease of maintaining the desired conditions.

Empirical expressions for EMC under various sets of drying air conditions have been reported by a number of investigators such as Day and Nelson (1965), Chung and Pfost (1967), Strohman and Yoerger (1967) Chen and Clayton (1971) and Thompson (1972). Pfost *et al.* (1976) compiled available data on equilibrium moisture content and tested the prediction accuracy of various models. They concluded that the three parameter equation of Henderson-Thompson (Thompson,1972) and Chung-Pfost (Chung and Pfost, 1967) give almost the same results as some of the more complex equations.

The investigator should refer to ASAE standards (D 2454) and select the most appropriate equation Hukill (1947) developed a number of dimension less curves and an equation for predicting the moisture content at any depth in a grain mass at any specified time. Equilibrium models, which assume that air and grain attain equilibrium at each layer have been developed by Bloom and shove (1971) and Thompson (1972). These models can be effectively solved using computers. The simultaneous heat and mass transfer approach has been used to develop partial differential models (Brooker *et al.*, 1974). These models include heat and mass-transfer coefficients, air and grain densities, specific-heat and thin layer equations. If sufficient data is available, these models can be employed.

Thompson *et al.* (1972) developed a deep-bed model based on heat and mass balance approach. The model considers the approach that a deep bed consists of a number of thin layers placed one over the other. The model uses 'equivalent temperature' values between grain and air before the exchange takes place. The rate of moisture exchange is calculated using an empirical thin layer drying equation.

Some other thin layer drying models include those of Thompson (1968), O' Callaghan *et al.* (1971), Wang and Singh (1978), Misra and Brooker (1980), White *et al.* (1981), Kulshrestha, Gupta and Singh (1987), Duggal, Gupta and Singh (1988) etc.

For equilibrium moisture content prediction, appropriate EMC model should be used. Some selected EMC models are given in ASAE (2001).

Kulsherestha (1983) worked on drying behavior of paddy (UPRI–8217). He considered 8 levels of inlet air temperature, 30, 35, 40, 45,50, 55, 60 and 65 °C and conducted experiments in thin layer as well as deep bed drying. He came up with the following equation for thin layer drying:

$$MR = \exp(-k\,t^{0.62})$$

where,

$$k = 4066.96 \exp(-3543.44/T)$$

He proposed the following empirical equation for relationship between apparent rate of paddy drying (per cent db/min) and drying air temperature(°C):

$$(Mo-M)/t = PT-Q$$

where,

$$P = 0.01049 \exp(-0.01026\,t)$$

$$Q = 0.21792 \exp(-0.01596\,t)$$

Time required (minutes) to dry paddy from harvest to safe storage moisture level could be correlated to drying air temperature (°C) by the following relationship:

$$t = 2748318.97\,T^{-2.7673}$$

Kulshrestha *et al.* (1990) tested the exponential model, generalized exponential model, Page's model, Wang and Singh model, Thompson's thin layer model and diffusion model and found that Page's model best described the experimental data in thin layer paddy drying. The model is of the form:

$$MR= \exp(-k\,t^{0.62})$$

where,

$$k = 4089 \exp(-3543/T)$$

T is absolute temperature in K and t is time in minutes.

Drying behaviour of 50 cm deep bed of paddy was studied (Kulshrestha, 1983) on a vertically stacked tray dryer having 10 trays and a provision to measure exit air temperature of each tray. Bulk drying behavior of the bed indicated a zero-drying period for low velocities (1, 2 and 4 m/min). Extent of moisture translocation increased with increase in air temperature or velocity. Simulated drying curves did not predict the zero drying period Simulated drying rates were, in general, higher than actual rates with a bed average moisture prediction error less than 1.73 per cent (db). Prediction error in moisture content for high velocity drying runs was less than 3.5 per cent for 90 per cent of data points with the average error of 1.25 per cent (db). Data are low air velocity being 5 per cent and 2.133 per cent (db) respectively.

Kulshrestha *et al.* (1987) examined the validity of Thompson's model in predicting deep bed drying behavior of paddy under non-equilibrium conditions. They concluded that Thompson's model predicts the moisture and temperature profiles of deep bed paddy drying under non-equilibrium conditions but errors as large as 2.26 per cent (db) and 4.84°C can be expected. The prediction errors were smaller at higher air velocities and lower drying temperatures.

Sutar (2003) studied the osmo-fluidized bed drying characteristics of onions in terms of effect of salt concentration (5, 12.5 and 20 per cent) and osmotic solution temperature (28, 43 and 58 °C) on moisture loss and solid gain during osmosis. The pre-treatment solution was of 0.2 per cent KMS. A 1:5 solid to solution ratio, 100 shakes per minute and 4 mm onion slices were used in the experiments. Twelve time intervals (5,10,15, 20, 25, 30, 49, 50, 60, 70,80 and 90 minutes) were used for studying the moisture loss and solid gain kinetics. It was observed that osmosis for 1 h at 20 per cent salt solution at room temperature (28°C) gave a moisture loss of 22.31 per cent of initial moisture and solid gain of 50.15 per cent of initial solids and this was the osmotic pretreatment before fluidized bed drying. The fluidized bed drying was carried out at four temperatures (40, 50, 60 and 70°C) and quality parameters namely rehydration ratio, dehydration ratio, optical density, reducing sugars, total sugars and sensory evaluation were carried out at each temperature. On the basis of these parameters, 70°C was selected as the fluidized bed temperature. Azura *et al.* (1992) model of the following form could describe the data well The model constants are given in Table 7.1.

$$ML_\theta = S_1\theta(ML_\infty)/(1+S_1\theta)$$

where,

ML_θ = Moisture loss fraction at any time,

q ML_∞ = Moisture loss fraction at any equilibrium

θ = Time, min

S_1 = Constant related to the rate of water diffusion out from product, min^{-1} and

$$SG_\theta = S_2\theta(SG_\infty)/(1+S_2\theta)$$

where,

SG_θ = Solid gain fraction at any time, θ

SG_∞ = Solid gain fraction at equilibrium

S_2 = Constant related to the rate of solids diffusion in the product, min^{-1}

The drying behavior of osmosed onion rings in fluidized bed could be described by Page's model. Sutar (2003) also proposed the following models for combined effect of temperature, salt concentration, and time of osmosis on moisture loss and salt gain as:

$$ML = 0.2856\, T^{0.5355}\, Co^{0.5328}\, \theta^{0.21079}$$

$$SG = 0.00781\, T^{0.7089}\, Co^{1.5523}\, \theta^{0.4558}$$

He used the following form of Page's model to describe the fluidized bed behavior of osmosed onions:

$$MR = \exp(-k\theta^n)$$

Table 7.1: Osmotic Model Constants for Moisture Loss and Solid Gain

Salt Concentration, %	*Temperature, °C*		
	28	*43*	*58*
		ML_{∞} (% of initial moisture)	
5	10.2541	13.0012	16.8808
12	15.9642	19.4363	21.3352
20	24.3546	27.7054	32.9500
		S_1 Min^{-1}	
5	0.1976	0.1895	0.1978
12	0.1207	0.1428	0.1468
20	0.1818	0.1728	0.3152
		SG_{∞} (% of initial solid)	
5	7.2569	9.3327	11.2942
12	45.6475	54.9692	58.8963
20	64.2508	100.2406	145.6240
		S_2 Min^{-1}	
5	0.1228	0.1084	0.0772
12	0.0330	0.0306	0.0550
20	0.0592	0.0398	0.0612

The constants of Page's model as worked out by Sutar (2003) are given in Table 7.2. He concluded that osmo-fluidized bed drying of onions is a potential alternate method of drying capable of yielding good quality finished product within 1 h and 49 minutes as compared to several hours in conventional drying method

Garg (1990) studied the drying behavior of lentils at four temperatures (50–80 °C) and four air velocities (3-60 m/min). The drying moisture range was from approximately 25 per cent (db) to 9 per cent (db). He also tested the validity of exponential and Page's models and found that page's model described the drying behavior of lentils quite closely. He found the following form of the Page's model suitable for lentils:

$$MR = \exp(-kt^{0.916})$$

and

$$k = 2387.6121 \exp(-3817.9338/T)$$

Table 7.2: Constants of Page's Model for Fludized Bed Drying of Onions

Temperature, °C	*K*	*n*
40	0.2445	0.5935
50	0.1845	0.7304
60	0.2499	0.6874
70	0.2988	0.7523

Source: Sutar, 2003.

Kanawade (1990) studied the fluidized bed behavior and quality of peas in relation to drying air temperatures of 60, 70, 80 and 90 °C Samples were pricked once in a hand operated pricking machine and blanched in hot water (90°C) for 2 minutes. He concluded that blanching had no effect on drying rate but pricking increased the drying significantly. The time required to dry the peas in fluidized bed varied from 110 to 310 min to dry from the initial to about 7 per cent moisture. The exponential type model with drying constant related to temperature by Arrhenius type relationship gave an excellent fit to the experimental data. The qualities of dried peas were influenced by drying air temperature, blanching as well as pricking. However, the effect of pricking was more than blanching. The drying air temperature of 70-80 °C with pricking and blanching were found to be optimum conditions for fluidized bed drying of peas.

Kar and Gupta (2001) studied the osmotic dehydration characteristics of button mushrooms quarters at 15 per cent salt solution. The sample to solution ratio was 1:4,1:6 and 1:8. They developed the following models to describe the dehydration behavior:

$$SC = a\ t^{b}\ T\ c\ R^{d}$$

where,

SC= Salt change, per cent

t– Time, minutes

T– Temperature, °C

R= Solution to sample ratio

The values of a,b,c, and d were 0.156, 0.373, 0.399 and 0.305 respectively.

For Solid loss:

$$MS= a + b \log t+ c \log T + d \log R$$

where,

MS stands for Mushroom solids loss.

The values of a,b,c, and d were -0.589, 0.184, 0.218 and 0.283 respectively.

Kar and Gupta (2003) tested the validity of exponential and Page's models in describing air drying data and found page's model to be more appropriate

Satter (1987) studied thin layer drying characteristics of pigeon pea (UPAS 120) grain. The grains were dried from 25 per cent to 8 per cent. The moisture content of 8 per cent was found the best for milling. The highest milling efficiency of 93.96 per cent was obtained with 80 °C drying air temperature followed by 3 h of tempering. At lower temperatures the milling efficiency was lower. A maximum of 9215 per cent efficiency could be obtained with sun-drying.

Anonymous (1987) found that the following model depicted the effect of flue gas velocity on the rate of drying of bagasse.

$$dM/dt= aV+ b$$

where,

dM/dt = Rate of drying of bagasse, per cent (db)/min

v = Flue gas velocity, m/min

a, b = Constants, depending upon initial moisture content of bagasse and temperature of drying air

Constant "a" varied from 0.00436 to 0.01474 and "b" from 0.0961 to 0.2965 depending upon the initial moisture content and temperature of drying air.

Duggal, Gupta and Singh (1988) fitted exponential and Page's models to the experimental data and came up with the following models:

$$MR= \exp (-kt)$$

where,

$$k = \exp(-3167.63/T), \text{ and}$$

$$MR = \exp(-kt^{1.14})$$

where,

k = 1474.13 exp(-3530.2/T)

T = Absolute temp, K

Singh (1991) studied the effect of drying conditions on textural characteristics of re-hydrated potato cubes. The drying conditions were drying air temperature of 40–80 °C and air velocity of 100–180 m/min. He concluded that a combined temperature-velocity model of the following form could describe the experimental data with a correlation coefficient of more than 0.85.

$$\text{Property} = bo + b_1 T + b_2 v + b_3 T^2 + b_4 v^2 + b_5 Tv + b_6 T^2 v + b_7 Tv^2 + b_8 T^3 + b_9 v^3 + b_{10} T^{-1} + b_{11} v^{-1}$$

where,

bo ...b_{11} are constants. The property means re-hydration ratio or hardness cohesiveness or springiness or adhesiveness or chewiness etc.

Goyal (2002) conducted two stage drying process for potato cubes of 1.0, 1.5 and 2.0 cm using air temperatures of 143.2 to 176.8 °C for 10 minutes and then 61.6 to 78.4 °C till the moisture reached 6.5 per cent. The cubes were blanched using KMS and $CaCl_2$ solution prior to drying. He reported that about half the moisture loss takes place in high temperature drying. The moisture ratio at any time could be described by Pages model.

References

ASAE. 2001. ASAE Year Book. Standard D 245.5. St. Joseph. American Society of Agricultural Engineers.

Anonymous. 1987. Optimum design of a bagasse dryer using waste heat from flue gases- final project report. G. B. Pant University of Ag. and Tech., Pantnagar.

Azura, F.,R., Cortes, H.S., Garcia, and C. I. Berstian. 1992. Kinetic model for osmotic dehydration and its relationship with Fick's Second law. *Int. J. Food Sc. and Technol*. 27,409-418.

Bloome, P.D. and Shove, G.C. 1971. Near equilibrium simulation of shelled corn drying. *Trans. ASAE* 14, 709-712.

Brooker, D.B. Bakker-Arkema, F.W. and Hall,C.W. 1974. Drying cereal grains. Connecticut: The AVI publishing Co.

Chen, C.S. and Clayton, J.T. 1971. The effect of temperature on sorption isotherms of biological materials. *Trans. ASAE* 14, 927-929.

Chen, C.S. and Johnson, W.H. 1969. Kinetics of moisture movement in hygroscopic materials-theoretical considerations of drying phenomenon. *Trans. ASAE* 12,109-113.

Chu, S.T. and Hustrulid, A. 1968. Numerical solution of diffusion equations. *Trans. ASAE* 11: 705-710.

Chung, D.S. and Pfost, H.B. 1967. Adsorption and desorption of water vapor by cereal grains and their products. *Trans. ASAE* 10, 549-557.

Day,D.L. and Nelson, G.L. 1965. Desorption isotherms for wheat. *Trans. ASAE* 8, 293-296.

Duggal, A., Gupta D.K, and Singh, B.P.N. 1988. Drying characteristics of sugar cane bagasse. *J. IE (I) AG* 68, 96-98.

Flood, C.A., Sabah, M.A., Meeker, D. and Peart, R.M. 1972. Simulation of a natural air corn drying system. *Trans. ASAE* !5, 156-159,162.

Garg, S. 1990. Thin layer drying characteristics of lentils. Unpublished M. Tech. Thesis. G. B. Pant University of Ag. and Tech.., Pantnagar.

Goyal. N. 2002. Optimization of two stage drying process and storage behavior of dehydrated potato cubes. Unpublished M. Tech.. Thesis. G. B. Pant University of Ag. and Tech., Pantnagar.

Henderson, J.M. and Henderson, S.M. 1968. A computational procedure for deep bed drying analysis. *J. Ag. Eng. Res* 13, 87-95.

Henderson, J.M. and Pabis, S. 1961. Grain drying theory I. Temperature effect on drying constant. *J. Ag. Eng. Res* 6, 21-26.

Hukill, W.V. 1947. Basic principles of drying corn and grain sorghum. *Ag Eng* 28, 335-338, 340.

Hussain, A., Chen, C.S. Clayton, J.T. and Whitney, L.F. 1972. Mathematical simulation of heat and mass transfer in high moisture foods. *Trans. ASAE* 15, 732-736.

Hustrulid, A. and Flikke, A.W. 1959. Theoretical drying curves for shelled corn. *Trans. ASAE* 2,112-114.

Kachru, R.P., Ojha, T.P. and Kurup, G.T. 1970. Drying characteristics of Indian paddy varieties. *J. Ag. Eng* 7,16-23.

Kanawade. V.L. (1990). Fluidized bed drying of peas. unpublished M. Tech. Thesis. G. B. Pant University of Ag. and Tech., Pantnagar

Kar, A., and Gupta, D.K. 2001. Osmotic Dehydration characteristics of button mushrooms. *J Food Sc. and Technol* 38, 352-357.

Kar, A., and Gupta., D.K. 2003. Air drying of osmosed button mushrooms. *J Food Sc. and Technol* 40, 29-31.

Kulshrestha. M. 1983. Simulation of drying behavior of paddy. Unpublished M. Tech. Thesis. G.B. Pant University of Ag. and Tech., Pantnagar.

Kulshrestha, M., Gupta, D.K. and Singh, B.P.N. 1987. Deep bed paddy drying under non-equilibrium conditions- Validity of Thompson's model. *J. IE (I) AG* 68, 1-4.

Kulshrestha, M., Gupta, D.K. and. Singh, B.P.N. 1990. Diffusion against empirical models for thin layer paddy drying. *J. IE (I)AG* 71, 28-30.

Misra, M.K. and Brooker, D.B. 1980. Thin layer drying and rewetting equations for shelled yellow corn. *Trans. ASAE* 23, 1254-1260.

Morey, R.V., Keener, H.M., Thompson, T. L. White, G.M. and. Bakker-Arkema, F.W. 1978. The present status of grain drying simulation. ASAE Paper No. 77-3013, St. Joseph. American Society of Agricultural Engineers.

O'Callaghan, J.R., Menzies, D.J., and Bailey, P.H. 1971. Digital simulation of agricultural dryers performance. *J. Ag. Eng. Res* 16, 223-244.

Overhults, D.G. White, G.M., Hamilton, H.E. and Ross, I. J. 1973. Drying Soybeans with heated air. *Trans. ASAE* 16, 112-113.

Pabis, S. and Henderson, S.M. 1962. Grain drying theory III. The air-grain temperature relationship. *J. Ag. Eng. Res* 2, 21-28.

Page, G.E. 1949. Factors influencing the maximum rates of air drying shelled corn in thin layers. Unpublished M. Sc. (ME) thesis. Purdue University, USA.

Satter, M.A. 1987. Drying and its effect on milling of pigeonpea grain (Cajanus Cajan L.). Unpublished M. Tech. Thesis. G. B. Pant University of Ag. and Tech., Pantnagar.

Sharaf-Eldeen, Y. I. Blaisdell, J.L. and Hamdy, M.Y. 1980. A model for ear corn drying. *Trans. ASAE* 23, 1261-1265,1271.

Sharma, A.D., Kunze, O.R., and Tolley, H.D. 1982. Rough rice drying as a two compartment model. *Trans. ASAE* 25, 221-224.

Simmonds, W.H.C., Ward F.T. and McEwen, W. 1953. The drying of wheat- Part I. 'The mechanism of drying' *Trans. Inst of Chem. Eng* 31, 266-277.

Singh, V.P. 1991. Effect of drying conditions on textural characteristics of rehydrated potato cubes. Unpublished M. Tech. Thesis. G. B. Pant University of Ag. and Tech., Pantnagar.

Spencer, H.M. 1926. Laboratory method for maintaining constant humidity. *International critical tables* 1,87.

Strohman, R.D. and Yoerger, R.R. 1967. A new equilibrium moisture content equation. *Trans. ASAE* 10, 675-677.

Sutar, P.P. 2003. Osmo-fluidised bed drying of onions. Unpublished M. Tech. Thesis. G. B. Pant University of Ag. and Tech., Pantnagar.

Thompson, T.L. (1972). Temporary storage of high moisture shelled corn using continuous aeration. *Trans. ASAE* 15, 333-337.

Thompson, T.L., Peart, R.M. and Foster, G.H. 1968. Mathematical simulation of corn drying- A New Model. *Trans. ASAE* 11, 582-586.

Wang, C.Y. and Singh, R.P. 1978. A single layer drying equation for rough rice. ASAE paper no. 78-3001. St. Joseph. American Society of Agricultural Engineers.

Watson, E.L. and Bhargava, V.K. 1974. Thin layer drying studies on wheat. *Can. J. Ag. Eng* 16, 18-22.

White, G.M., Bridges, T.C, Loewer, O.J.and Ross, E. J. 1981. Thin layer drying model for soybean. *Trans.ASAE* 24, 1643-1646.

White, G.M., Ross, I.J. and Westerman, P.W. 1973. Drying Rate and quality of white shelled corn as influenced by dew point temperature. *Trans. ASAE* 16, 118-120.

Whiteker, T.B., Barre, H.J. and Hamdy, M.Y. 1969. Theoretical and experimental studies in spherical bodies with a variable diffusion constant. *Trans. ASAE* 12, 668-672.

Chapter 8

Storage of Food Grains

Sanjay Kumar Jain and R.C. Verma

Department of Processing and Food Engineering,
College of Technology and Engineering, Udaipur – 313 001

The technological advancement in agriculture and its allied fields is greatly responsible for higher food production in India. A record 212 million tonnes of food grains were produced in 2002-03 as compared to 55 million tonnes in 1950-51 in the country. The demand for food for the ever-increasing population is continuous but production is limited due to specific seasons. Food Corporation of India and other government agencies procure and handle a very small percentage of this huge production and most of it remains with the farmers and traders. According to one estimate, about 70 per cent of the total food production is retained in the rural areas to meet farmers' food and seed requirements and the rest enters the urban market channel and public distribution system. Therefore, the food grains are required to be stored for some period. The period may vary from a few weeks to a few years depending upon the requirements. The Indian picture reveals that a sizeable quantity of about 10-20 per cent of the retained food grains is lost due to improper handling and inadequate storage conditions.

Grain is a living biological material and therefore storage structures should be suitably designed to protect it from the ravages of pests, rodents and micro-organisms. However, Food and Agricultural Organization has estimated very high losses in food grains in developing countries under tropical climates. Although the grain losses occur at different stages beginning from the farmer's field till it reaches the final consumer but the storage losses account for a major portion. These losses can be reduced to a considerable extent if proper care is taken. Unfortunately farmers are not adequately equipped and trained to store grains safely for more than a few weeks. Therefore, they prefer to sell their produce immediately after threshing the crops. Even if they store the grains, they have to face numerous problems during storage and huge quantity of food grains is spoiled. There are many factors responsible for grain spoilage, however, pests and rodents are the primary factors especially in the regions of high humidity and high temperature. Faulty design, poor maintenance and limited knowledge of the farmers concerning storage technology has greatly added to the spoilage of the food grains during storage. Sometimes the food grains if not properly stored, may be edible but they may lose essential vitamins and minerals and hence their nutritive value. Adding further more to adverse effects, food grains may become toxic even.

The storage structures used in different parts of the country are of numerous types and there are many variations in their materials, constructional methods and capacities. These storage structures are not technically sound and modifications are highly desirable. The time has come to minimize food grain losses and wastage to feed the ever-increasing population and this is particularly more important in the Indian context where cultivable land is decreasing at a fast pace. Taking into consideration major features of a good storage, it should be moisture proof, rodent and termite proof, equipped with easy loading and emptying mechanism, cheap and include the local materials in construction. Also it should not cause the loss of structure of grain by dehydration (wilting, shriveling *etc.*) and have low thermal conductivity and provide air tightness. However, all of the above parameters are not fulfilled in the rural farm storage causing high loss of food grains.

Losses of Good Grain

The qualitative and quantitative losses in food grains may occur separately or together. The chemical changes occurring in protein and carbohydrates and the contamination taking place due to chemical toxins, insects' fragments, rodents' urine and excreta are mainly responsible for the loss in quality of the food grain during storage. During storage, both the losses occur simultaneously. The seed loss is another major loss and many factors are responsible for this. The viability of seeds may reduce if seed coat is damaged during filling and emptying the storage structures. The chemicals used to control the insect infestation also affect seed germination. The viability is also affected by the development of micro-organisms. The chemicals cause abnormal growth of root and shoot, reduction in vigour and other undesirable physiological changes. The factors influencing the storage losses are moisture, temperature, insects and rodents and conditions of the storage.

Grain Moisture

The moisture content is the single most important factor responsible for the deterioration in the quality of the food grains during storage. It is often said that reduction in 1 per cent of moisture content doubles the shelf-life. The amount of moisture in grain determines whether mould damage can occur. If there is sufficient moisture for mold growth, the temperature determines which mold growth can occur and how fast it can develop. In any case, higher the moisture and temperature of the food grains, the faster is the growth and more rapid is the spoilage of grain. However, when the moisture content is low, all mold growth stops even at favourable temperature. There is a certain limit of grain moisture of different commodities in relation to prevailing conditions of environment that should be kept in mind before storing the grains. The moisture to which grain will dry is determined by the relative humidity of the surrounding atmosphere. This property of grains is referred as equilibrium moisture content. The maximum level of moisture content for safe storage is given in Table 8.1.

Table 8.1: Equilibrium Moisture Content (EMC) of Various Food Grains at 70 per cent RH

Commodity	*Paddy*	*Rice*	*Maize*	*Wheat*	*Sorghum*	*Cotton seed*	*Groundnut (Shelled)*	*Copra*
EMC	15	13	13.5	13.5	13.5	10	7	7

The moisture is also an important factor affecting the activity of insects. It has been found that when the moisture content is as low as 9 percent in shelled corn and wheat, most insects become inactive. While such low moisture content is neither readily obtainable in current practices, nor desirable because of shrinkage. Experience shows that drier the grain, lesser the insect hazard. Table 8.2 shows the safe storage moisture for storage period of one and five years for various food grains. These limits of moisture have been fairly well established through research and experience.

Table 8.2: Safe Storage Moisture for Various Food Grains

Time	*Moisture Required for Safe Storage per cent (wb)*						
	Barley	*Corn (Maize)*	*Oats*	*Rice*	*Rye*	*Sorghum*	*Wheat*
1-years	13	13	14	12-14	13	12-13	13-14
5-years	11	10-11	11	10-12	11	10-11	11-12

Temperature

The temperature controls the activities of the micro-organisms and its effect is also correlated to the moisture present in the grain. When grain in storage is cool, there is little likelihood of spoilage. Low temperature offsets the effects of high moisture, particularly as it affects the development of molds and insects. One of the greatest benefits of moving and turning grain, so widely used in elevators, is cooling. This, however, can be accomplished more effectively by aeration, which is used to cool and hold damp grain in storage for weeks or even months. The respiration rate of grains is low at low temperature. Under favourable conditions, sometimes the respiration rate reaches so high that the germination occurs. These chances are even more in the freshly harvested grains with high moisture, if stored without aeration. The grains contain initial internal infestation as well as field and storage fungi. Since the infestation and infection by storage pests and organisms are inherent, the atmosphere inside the storage should be capable of depletion of oxygen and moisture migration. The temperature gradient and moisture migration are generally present in the metallic structure with high thermal conductivity. This may induce the condensation of moisture within the bulk. This condensed moisture increases the higher moisture pockets. It may be responsible for high rate of respiration, mould growth and threatened germination. If a light roof coloured with white paint over the top (which reflects the solar heat) is provided then the heat transfer can be minimized. The insects accelerate their growth up to 42°C but fungi have tolerance limit from 20°C to a maximum 63°C. Therefore this environment has to be avoided for safe storage requirements.

Fungi, Mites, Insects and Rodents

The micro-organisms are omnipresent. They may come from the soil, growing plant, surface of container and the air. There are two types of micro-organisms. The budding and fission fungi are the most dangerous and extremely harmful for the grain as they cannot be seen with naked eyes and their harmful influence spreads very quickly. They consume the soluble constituents of the food tissue and may generate organic compounds that make food unpalatable. On the other hand, the presence of rodents is conspicuous by the holes. Most storage fungi grow at faster rate in the temperature range of 20 to 40°C under high humidity conditions. However, growth of some of the species is restricted under these conditions. Some fungi produce chemicals, which are toxic to human and domestic animals. The growth of these fungi is more in the coastal areas of the country where high humidity and favourable temperature exists.

Mites are extremely small having a shining translucent body and legs. They vary in colour from pale straw to dark reddish brown. Mites are usually seen if they are large in numbers and are visible as dust on the surface of the bags.

The damage from the insects is the principal loss in stored grains in the coastal areas of the country where the climate is warm and humid. The insects may also cause problem in low temperature areas if the moisture content is high and there are sources of infestation. The undesirable insects/pests causing deterioration and loss to stored grains are mainly beetles and moths. They do not breed successfully if the relative humidity of the environment is less than 40 per cent. Cereals are vulnerable to rodent attacks and damage of this effect is the maximum at the village level farm storage. Rodents carry diseases transmissible to the human beings and are also a source of contamination within the stored grains.

Quality Before Storing the Grain

The objective of grain storage is to maintain the qualities and characteristics that the food material possesses after harvesting and drying. That means the viability should be preserved for seeds, milling and baking qualities for the grains stored for processing and nutritive qualities for the grains fed to animals. In any case, the quality of the grains before storing in the bins plays an important role. During storage, the quality of food material cannot be improved. If infested grain is stored in the bin, there is high probability that the infestation will grow. At the same time infested bins also accelerate the contamination of the grain during storage. Many of the problems about poor quality of the food grains are not due to improper storage but due to early harvesting at high moisture level and rapid drying. It is therefore, desirable to store uncontaminated clean grain in insect-free and sound storage structures of proven type.

Grain Storage Structures

The preliminary storage of food grains begins at the farm itself after the crop is harvested. In India, freshly harvested crops are heaped on the threshing floor and remain there for quite sometime before threshing. However, it should not be kept for more than a day or two without drying. However, these heaps can be stored for longer period if harvesting conditions coincide with dry weather, which also helps in drying to some extent. There after, the threshed grain is stored either indoor or outdoor structures. The indoor storage structures are employed by the small and marginal farmers with the main objective of storing the grain both for food and seed. Outdoor structures are seldom used by the small farmers. Storage can also be distinguished as (a) underground and (b) above ground.

Underground Storage Structures

These types of storage structures found their place in low water table region especially in the dry parts of Karnataka, Uttar Pradesh, Bihar, Madhya Pradesh, Maharashtra, and Andhra Pradesh. The safety and ease in maintenance are the prime factors responsible for their popularity. They are rectangular or conical in shape and have narrow mouth to fill the grains. These types of structures are easy to construct as only pits are to be dug-up and lined with straw to absorb the moisture. After filling the food grains, the inlet is covered with the stone slab. Infestation is generally low due to depletion of the oxygen but seepage of moisture may result in the mold growth. The rodents are also a problem in these types of structures. The loading of the food grains and fumigation during storage is a problem in these types of structures. The peripheral layer of grains may sometimes get spoiled due to seepage from the walls or from the bottom of the structure. The grains stored in these structures may become dull in colour and mushy in odour and may not be suitable for human consumption. The viability of seed is also affected. The Central Food Technology Research Institute, Mysore has improved these type of structures by giving a polyethylene lining on the walls, periphery and at the floor which reduces the ingress of the moisture.

Above-ground Storage Structures

Bag Storage

Generally the capacity of gunny bags is 50 or 100kg. The traders and processors store grains in gunny bags as they are very convenient in handling. The gunny bags can be placed in any room and therefore these types of structures are simple in construction and low in cost. Traders store grains for trading purposes and they seldom worry about the quality of food. The problem of deterioration exists in all types of storage structures, however the degree of damage varies. Some details of various storage structures are given in Table 8.3.

Bukhari

Bukhari type structures of varying shapes, size and material are used in different parts of Assam, West Bengal, Orissa, Bihar and Madhya Pradesh. The structure is made of split bamboo-strips with a bottom either made of bamboo or wooden planks. They are mostly cylindrical in shape. The structure is generally used for storing seeds like wheat, gram, maize and sorghum. The structure is also used for storing wheat straw for cattle feed. Generally 1.0-1.5 tonnes of seed may be stored by the farmers although the space available can accommodate a lot more than this quantity. A 15 cm thick layer of wheat straw is first spread at the bottom before placing the seed bags in stack. Finally the bags are covered with straw till they remain in storage. An outlet is also provided at the bottom of the structure. The outlet in the lower portion is not necessarily for taking out grains but it is also helpful in overall cleaning of structure. The general defects like inadequate height of the platform, unsuitability of structure for fumigation, not being moisture and fire proof and vulnerable to insects and rodents have been minimized in the improved Bukhari type structures.

Table 8.3: Details of Various Food Grain Storage Structures

Sl.No.	*Type of Bin*	*Material of Construction*	*Capacity, kg*	*Expected Life, Years*	*Suitability for Fumigation*
1	Untreated bamboo bin	Bamboo strips	200	1	No
2	Clay and cow dung plastered bamboo bin	Bamboo clay and cow dung	200	3	No
3	Bamboo cement bin	Bamboo, cement and sand	200	10	Yes
4	Bamboo linseed oil and ash bin	Bamboo, linseed oil, ash and cow dung	200	12	Yes
5	Bamboo reinforced concrete bin	Bamboo cement concrete	325	35	Yes
6	Ferro cement bin	Wire-mesh, cement, sand and mortar	150	40	Yes
7	Steel reinforced bin	Steel reinforced cement concrete	325	40	Yes
8	RCC ring bin	Steel reinforced cement concrete	1000	40	Yes
9	*Pucca* kothi	Brick, cement and mortar	1000	40	Yes
10	Metal bin	Galvanized Iron sheet	1000	35	Yes
11	Bomboo mud plastered polythene lined bin	Bamboo, clay, cow-dung, polythene	160	3	Yes
12	*Pusa* bin	*Kachha* bricks. Cow-dung, polythene	300	12	Yes

Oil and Coal Tar Drum Storage Structure

Abandoned drums of oil and coal tar are utilized by some villagers to keep sizeable quantity of their farm produce. These are readily and economically available in the towns and cities. The capacity of 220 litre drum is about 175 kg of grain. An improvement can be made by providing the lid half-open and using some metal primer after cleaning the container thoroughly. These bins are reasonably cheap, durable and suitable for grain storage. It would be desirable to empty the drum by carefully cutting the top portion of the cylinder so that drum could be later modified as storage structure.

Morai

Morai is exclusively made of paddy straw ropes. Sometimes mud is also used for plastering the bottom portion of the structure. There are different local names prevalent for the same structure. In Andhra Pradesh it is called *'Puri'* and *'Mora'* is the term used for the same in north Bihar and *'Oliya'* in Orissa. The holding capacity of the structure also differs from location to location and according to the need of the owner. In West Bengal, its capacity ranges from 2 to 5 tonnes. It is constructed over a hard and elevated surface utilizing 10-15cm diameter straw rope, which is wound to form a concentric circular wall. Simultaneous loading and winding operations are performed. Sometimes metallic base is provided as a part of Morai which is very useful to provide additional support to the structure. The brick masonry base constitutes the additional cost of structure. The structure is susceptible to fire and it is neither air-tight nor moisture proof. Insects and rodents can cause sizeable damage to the stored grains.

Large Scale Storage Structures

The storage and handling of food grains by various agencies like CWC (Central Warehousing Corporation), SWC (State Warehousing Corporation), FCI (Food Corporation of India) and Government Civil Supplies Department have assumed greater importance now a days. A number of godowns and ware-houses have been built to store the food grains. The following are few techniques generally used by these agencies for large handling of the grains:

Cover and Plinth

Nowadays Food Corporation of India has to handle large quantity of the grains for which it does not have adequate storage facilities. Therefore a new and innovative technology of sorage *i.e.,* cover and plinth (CAP) was introduced to overcome the lack of storage space temporarily. It has become a permanent system of storage in many ware-houses of FCI, CWC and SWC's. These structures are made up of heavy duty LDPE film. It is an improved structure for storing the grains in the open on the plinth which is essentially a wooden dunnage having damp and rat proof arrangements. The grain bags are stacked to standard size like any other commercial storage space. Previously each stack of grains was covered by polyethylene sheet or by the tarpaulins to protect the grains, but presently grain stacks are covered by the 1000 gauge LDPE sheet. Wheat, maize, gram, paddy and sorghum are generally stored in CAP storage facilities. CAP cover weighing 50 kg can store 150 to 200 metric tonnes of food grain for 1 to 1.5 years. Thus the CAP covers are the most economical of all the storage methods adopted.

Silo Storage

The bulk grain storage is now gaining popularity specially where grain is handled in a big quantity. The metallic and cement concrete structures have limitations regarding the diurnal temperature changes. The fungal attack is a big problem in these types of structures due to development of thermal gradient, moisture migration and localization of moisture pockets. Apart from this, due to 30 per cent void spaces present in the storage structures, these cannot provide the conditions to inhibit insect growth. Once the growth of insects and microbial activity starts in one pocket, it catalyzes the deteriorative activity from one focus to another. In order to overcome these difficulties, aeration arrangement or turning the grains is absolutely necessary. This requires pneumatic or other mechanical feeding and discharging along with aeration and fumigation devices. These portable equipment are economical only where minimum size of the storage is more than 5000 tonnes and turnover in a year is at least 3 times its capacity.

Pusa Bin

Indian Agricultural Research Institute, New Delhi has improved the conventional grain storage structures by inserting a LDPE film between two layers of mud walls. It is termed as pusa bin. The LDPE provides the moisture-proof arrangement and the mud walls keep the stored grains free from the effect of temperature changes. This bin has been successfully used for the various types of food grains and it has been reported that the seed stored in this type of storage structure retains its quality over a long period of time. The construction of bin is easy and cheap as compared to other scientific grain storage structures. The pusa bin provides damp proof structure as well as it limits the permeability of gases. It has low thermal conductivity and has resistance to rodent attack.

Chemical Control of Insects and Rodents

In India insects, micro-organisms (fungus) and rodents cause heavy losses to the food grain. Ravages of such pests are most noticeable in rural household, warehouses and on farmlands. In most cases, the use of pesticides is the only reliable method of destroying them and thus preventing these losses. Since, most of the rural based storage structures are not scientifically pest proof, the application of pesticides is highly advocated. There are a number of following factors, which must be kept in mind before planning the spray or fumigation of the stored grain both in large and small-scale storage systems:

1. Type of cereal being stored,
2. Type of pest present or likely to attack,
3. Type of storage structure being used, projected storage period,
4. Chemicals available and expertise to handle it and
5. Ultimate use of the grain either as seed or animal feed or for human consumption.

As far as possible, established chemicals should be used as fumigants for control of insects and insecticides to prevent re-infestation. Table 8.4 shows the list of some of the common insects, pests, and rodents found in stored grains.

Once the factors of infestation have been carefully looked upon, one may conclude that following definite characteristics are required for a fumigant:

1. A fumigant should be highly toxic for insects at all stages of life cycle.
2. It should have high vapour pressure and have no adverse reaction with the stored grains.
3. It should be non-inflammable and devoid of causing fainting.
4. It should not settle down immediately after application.
5. It should not affect the viability of treated seed and should have known antidote and treatment.
6. It should be indigenously manufactured and economical.

Table 8.4: Common Insects, Pests and Rodents of Stored Grains

Local Mame	*Scientific Name*
Rice weevil	*Sitophilus oryzae*
Granary weevil	*Sitophilis granarius*
Lesser grain borer	*Rhizopertha dominica*
Khapra beetle	*Trogoderma granarium*
Pulse beetle	*Collosobruchus spp*
Red flour beetle	*Tribolium cost anem*
Confused flour beetle	*Tribolium confusum*
Saw toothed grain beetle	*Oryzaephilus surinamensis*
Grain moth	*Sitotraga cerealella*
Rice moth	*Corcyra cerealella*
Fig moth	*Ephestia cautella*
House mouse	*Mus musculus*
Black rat	*Rattus rattus*
Brown rat or Norway rat	*Rattus poryegicus*
Lesser bandicoot	*Bandicoot bengal ensis*
Larger bandicoot	*Bandicoot indica*

The use of chemicals would be effective, only if applied in proper doze to cover particular area or volume. Table 8.5 shows the application rate of various chemicals. In addition to the chemical methods, farmers also use some traditional methods to destroy insects. Conditioning in the sun at a regular interval is one of them. Insects feel uneasy at a temperature above 40-44°C and try to runaway from the grain mass. However, conditioning in sun does not kill the eggs and larvae. *Neem* (*Azadirachta Indica Juss*) leaves and ash are commonly used in stored grains in Bihar and West Bengal. Sometime, onion bulbs of smaller sizes are placed in the stored grain. These admixtures work as repellants to insects. Wood-ash and chopped wheat straw are also used especially when grain is to be used for seed purpose. By adding straw or ash, insects are deprived of oxygen and also feel uneasy in crawling. Smoking is also used as one of the measures to check insect infestation

Table 8.5: Recommended Dosages of Widely Used Chemicals for Stored Grain

Chemical	*Dosages*
Malathion	Dust
	– 0.250 kg per 10 m^2 of surface area
	– 0.120 kg per 10 m^2 of surface area
	Spray
	– Surface treatment: 3 litre per 300 m^3 at an interval of three weeks
	– Air charging: 1litre per 300 m^3 at an interval of not less than seven days (two-three weeks)
	– Walls and alleyways: 3 litre per 100 m^3 once a week in order to control crawling insects
Aluminum phosphide	100 to 140 tablets per 100 m^3
Methyl bromide	1.5 to 2.2 kg per 100 m^3
Benezene hexochloride dust (BHC)	100 g of BHC containing 0.65 per cent Gamma isomers for 4 m^2 of surface
Carbon disulphide	2kg per 30 m^3 space holding about 20 tonnes of stored grains
Lindane (Gamma BHC)	225 g of 0.04 per cent lindane per 45 kg of grain
Pyrethrum	0.5 kg pyrethrum per 90 kg of maize

Code of Practices for Safe Storage of Grains

The traditional methods practiced by the farmers with regard to the storage of food grain and seed have evolved mainly on the basis of their past experience passed on to them by their forefathers. Such practices are not always scientific and may not be considered safe to be followed for preventing grain losses during storage. Based on the scientific principles and successful results obtained in the past, the following codes of practices are recommended to the small and marginal farmers by Indian Grain Management Research Institute. There may be different opinions for sequence of practices but above all, practices followed have the same objective and that is to store grain without loss. It is also pointed out that the extension workers, the traders and others involved in the grain storage programmes may also take note of the practices in making their programme a success.

Precautions for Storage

1. As far as possible, clean the grain prior to storage.
2. Dry the grain either in the sun or in the dryer till moisture is reduced to the safe limit.
3. If maize is to be stored, put kernels into the structure right after removing them from the cob. Maize should not be exposed to prevailing weather, once it has been dried.
4. Clean the storage structure and disinfect it before filling the grain.
5. Fill the structure up to the capacity carefully, preferably from the top.
6. Stir grain with a large stick so that the grain settles.
7. Add recommended doses of insecticide to the grain storage structures before closing the cover.
8. Close inlet and outlet covers immediately as flying insects may enter the stored grains.
9. Put mud or other binding material as the case may be on cracks and crevices. This does not allow the insects and pests to enter the grain or the structure.
10. Area around the structure must be cleaned to discourage rats hiding around. Rats do not like clean premises.
11. Inspect the structure regularly to detect cracks etc. Plaster the cracks after disinfecting.
12. Keep the outlet cover locked in the outdoor structure to avoid any risk of pilferage.
13. Spilled grains around the structure should be removed otherwise insects and rats can be attracted.
14. The cover should be firmly plastered after removing the grain or seed from the structure.
15. No grain should be left in the structure while emptying or, otherwise it can attract undesirable insects and pests.
16. Sweep all the debris and dust etc. and disinfect the structure.
17. The structure should be thoroughly cleaned and made free from insects before filling the next lot. A small fire

sufficient to create smoke is advisable in case the structure is fire proof.

18. Once the structure is repaired, cleaned and disinfected, it is ready for the next filling.
19. It is always desirable to consult the local agencies regarding the use of chemicals and improved techniques for grain storage.

Code of Practices for Safe Storage of Pulses

Pulses are the main component of the Indian diet and increment in the production of pulses has also added to the problem of their safe storage. It is always required to maintain food quality in storage and pulses are not an exception. The main pulses grown in India are black gram, bengal gram, green gram, horse gram, pea, moth, beans, pigeon pea and lentil. A considerable portion of the pulse production goes waste due to unscientific method of storage mainly due to insect infestation, moisture, rodent and bird menace. The proper care during the storage has, therefore, become essential to check this ever-increasing loss, to maintain the nutritive value of pulses and to ensure good quality protein. The following are a few rules for proper storage of pulses:

Drying and Cleaning

Dry and clean the pulses before storing. The high moisture pulses generate heat during storage, which may develop musty odour, discolouration and lump formation. This may also attract the insects and ultimately the stock may become unfit for human consumption. In order to overcome these problems, the pulses should be dried up to the level of 11-12 per cent moisture or even less. The pulses may be dried to this moisture under the sun if favourable conditions prevail or they may be dried with the help of mechanical dryers.

Use of Dunnage

If gunny bags are used for storage of pulses, then it is always desirable to provide the dunnage to avoid the moisture damage to the pulses. Wooden crates, polyethylene sheets and bamboo mats are most suitable for this purpose as they prevent moisture migration from the ground.

Use of Scientific Storage Structure

It is always preferable to use metal bin or *pucca* kothi to store pulses. This eliminates the rat menace and insects can be easily controlled in these types of structures. Further, these also provide the effective shelter to pulses and prevent losses due to birds.

Fumigation

The main insect which attack the pulses are *Callosobruchus chinesis lin, C. maculatus Fb., Bruchus analis Fb., B. phaseoli Gyll.* and *Laria affinis Frol.* These beetles mainly attack black gram, bengal gram, green gram, horse gram, pigeon pea and lentil. The damaged pulses show round holes, which are the exit holes of the beetles. In heavily infested pulses, cotyledons are completely eaten by the larva leaving only the outer covering. *C. chinensis* prefers to attack green gram and rarely attacks black gram. *C. maculatus* attacks cow peas, small and large peas, beans, bengal gram, red gram, black and green gram, soybean. *B. analis* is found in association with *C. chinesis. Bruchus phaseoli* is a serious pest of cow-pea but also attacks other pulses. *Laria affinis* is commonly called pea bruchid because of its preference for small peas. Generally it attacks the crop when it is in the field. Generally the pulse beetles do not attack split pulses if they are sufficiently dry. Sorghum should not be stored adjacent to pulses because it is often found infested by pulse beetles. Sometimes split pulses, which are stored for a long time and show attack of *bruchid* area are also infested with rice moth *Corcyra cephalonica Staint*. The presence of this insect can be recognized by the webbing of small pieces of broken pulses and flour inside which the cater pillar feeds.

For controlling insect infestation in the store, fumigate the pulses with EDBA or EDBA + CT mixture ampules at the rate of 30 ml or 110 ml/tonne respectively or aluminum phosphide at the rate of 2-3 tablets (3g each) per tonne. The exposure period should not be less than 7 days. The receptacles in which fumigation is to be done should be air tight. For fumigation of pulses in bags, polyethylene sheet should be used to cover them. It is always safe to undertake fumigation under the supervision of a technical person.

Control of Rodents

Rodents are a serious problem and may cause considerable damage to pulses stored in gunny bags or in the bulk. The use of *pucca* kothi or metal bin is the alternate solution. The use of poisoning bait with anti-coagulants causes slow death without creating bait shyness amongst the rodents. The bait is generally prepared by mixing crushed grain (cereals/millets), edible oil, sugar or jaggery and anticoagulant rodenticide in the quantity of 450g, 10g, 15g and 25g respectively. The bait thus prepared is kept in shallow containers for a period of 3 weeks and the rats start dying after 5 to 7 days.

References

Brooker, D B, Arkema F W B and Hall, C W. 1978. Drying Cereal Grains. The AVI Publishing Company. Inc.

Jain, S.K and Verma R C. 2002. Grain Storage at a Glance. Lecture notes of ISTE/AICTE summer school on Information Technology in Context with Food Engineering during June, 10-23, 2002. PP: 128-137.

Sinha, R N and Muir. 1975. Grain Storage: Part of a System. The AVI Publishing Company Inc.

Chapter 9

Pre-Cooling of Horticultural Produce

Satish Kumar and Mahesh Kumar

Department of Processing and Food Engineering,
Punjab Agricultural University, Ludhiana – 141 004

Fruits and vegetables are inseparable part of human diet as they provide nutritionally important, complex chemical compounds called vitamins which cannot be synthesized in our body and have to be ingested from external sources. We need supply of fruits and vegetables throughout the year in recommended quantities. Unfortunately, the horticultural products are perishable as they continue to perform physiological functions (respiration, transpiration etc.) even after harvest which causes their deterioration. Pathological and physical degradative processes also spoil their quality. Chemical and enzymatic changes may cause tissue softening, pigment loss and off-flavours (Dennis, 1984). Thus in order to extend the shelf-life of perishable products we have to either retard these deteriorative processes by following suitable postharvest handling practices or inactivate these by employing food preservation techniques. These techniques are again influenced by the quality of the raw-product. The postharvest handling practices not only determine the quality of whole raw- product but also the quality of various end products.

Physiological changes that take place in fruits and vegetables during postharvest handling are influenced by the various intrinsic and extrinsic conditions. Careful examination of these conditions is essential for the application of proper remedial measures. Temperature of produce and the environment have been observed to be the major factors contributing to overall useful life of these plant materials. Consequently pre-cooling of horticultural produce to reduce field heat has been accepted as the single most important operation contributing in a big way to extend of shelf-life and quality of perishable produce.

Postharvest Physiology

Following activities of the living system influence the development and ripening of horticultural produce:

Respiration

It is basically an oxidation reaction wherein starch, sugars, organic acids and other food reserves are utilized and converted into complex molecules like carbon dioxide and water accompanied by release of energy (For each milligram of carbon dioxide produced by respiration, 2.55 calories of heat is generated. Since 1 calorie = 4.187 Joules, thus 1 milligram of carbon dioxide produced generates 10.67 Joules of energy.).

$$C_6H_{12}O_6 + 6O \longrightarrow 6CO_2 + 6H_2O + \text{ENERGY}$$

The CO_2 thus formed escapes, leading to weight-loss while the energy released tends to raise the temperature of the produce. The rate of respiration is governed by temperature. For every 10°C rise in temperature, the rate of respiration roughly doubles or triples (Table 9.1) as the rate of most chemical and biological reactions increases 2 to 3 times. Reid and Kofranek (1980) found that rates of respiration of roses and carnations increase 25 times between 0°C to 20°C. Winkler *et al.* (1974) observed that a reduction in temperature by about 10°C in grapes, halved the rate of respiration and doubled their keeping quality. Therefore, the rate of respiration is often considered as a good index of anticipated storage life of horticultural produce. Higher is the rate, shorter the life of the product and vice versa.

Table 9.1: Respiration Rates of Fruits and Vegetables, Expressed as Rates of Carbon Dioxide Production (mg/kg/h), at Various Temperatures

Sl.No.	*Commodity*	*Temperature*			
		0°C	*5°C*	*10°C*	*20°C*
1.	Apples, summer	3–6	5–11	14–20	20.41
2.	Apples, fall	2–4	5–7	7–10	15–25
3.	Bananas, green	–	–	–	33–35
4.	Bananas, ripening	–	–	21–39	33–142
5.	Beans, lima	10–30	20–36	–	133–179
6.	Beans, snap	20	35	58	130
7.	Bean sprouts	21–25	42	93–99	130
8.	Strawberries	12–18	16–23	49–95	102–196
9.	Broccoli	19–21	32–37	75–87	278–320
10.	Brussels sprouts	10–30	22–48	63–84	86–190
11.	Cabbage	4–6	9–12	17–19	28–49
12.	Carrots, topped	10–20	13–26	20–42	46–95
13.	Cauliflower	16–19	19–22	32–36	75–86
14.	Cherries, sweet	4–5	10–14	–	28–32
15.	Cherries, sour	6–13	13	–	39–50
16.	Grapefruit	–	–	7–9	13–26
17.	Lemons	–	–	11	19–25
18.	Oranges	2–5	4–7	6–9	22–34
19.	Cucumber	–	–	23–29	14–48
20.	Garlic	4–14	9–33	9–10	13–25
21.	Grapes, American	3	5	8	33
22.	Grapes, vinifera	1–2	3–6	8	–
23.	Kiwifruit	3	6	12	16–22
24.	Litchies	–	–	–	–
25.	Mangoes	–	10–22	–	75–151
26.	Cantaloupes	5–6	9–10	14–16	45–65
27.	Honey dew	–	3–5	7–[illegible]	20–27
28.	Water melons	–	3–4	6–9	17–25
29	Mushrooms	28–44	71	100	264–316

Contd...

Table 9.1–Contd...

Sl.No.	*Commodity*	*Temperature*			
		0°C	*5°C*	*10°C*	*20°C*
30.	Onion, dry	3	3–4	7–8	14–19
31.	Onion, green	10–32	17–39	36–62	79–178
32.	Okra	–	53–59	86–95	248–274
33.	Papaya	–	4–6	–	–
34.	Peaches	4–6	6–9	16	59–102
35.	Pears, Bartlett	3–7	5–10	8–21	30–70
36.	Peas, unshelled	30–47	55–76	68–117	245–361
37.	Peas, shelled	47–75	79–97	–	349–556
38.	Peppers, sweet	–	10	14	44
39.	Pineapples, mature green	–	2	4–7	19–29
40.	Plums	2–3	4–9	7–11	18–26
41.	Potatoes, immature	–	12	14–21	18–45
42.	Potatoes, mature	–	3–9	7–10	8–16
43.	Radish with tops	14–17	19–21	31–36	124–136
44.	Radish topped	3–9	6–13	15–16	44–58
45.	Spinach	19–22	35–58	82–138	172–287
46.	Squash, summer	12–13	14–19	34–36	85–97
47.	Sweet corn	30–51	43–83	104–120	268–311
48.	Tomatoes, mature green	–	5–8	12–18	28–41
49.	Tomatoes, ripening	–	–	13–16	24–44

Source: Hardenberg *et al.*, 1990.

Respiration quotient (RQ) defined as the ratio of volume of CO_2 released to the volume of O_2 absorbed in the respiration indicates the nature of respiratory process. For aerobic respiration, RQ value lies between 0.7- 12.3 which can further be divided on the basis of constituents being oxidised *i.e.* for carbohydrates RQ = 1, for lipids RQ < 1 and for organic acids RQ > 1. On the other hand under anaerobic conditions there is incomplete oxidation and the RQ values are more than 1.3.

Factors Affecting Respiration

Temperature

Temperature is the most important environmental factor affecting the rates of biological reactions including respiration. The influence of temperature on the degradative reaction is indicated by Q_{10} value (temperature quotient), which represents the ratio of change in the rate of reaction to 10°C change in temperature. It has been observed that average Q_{10} values indicating the rates of reaction are 1.0 to 3.0 at 10°C, 7.5 at 20°C, 15.0 at 30°C and 22.5 at 40°C (Kader, 1987). The Q_{10} values for rate of respiration decline at 50-55°C as the tissue nears its thermal death point.

Ethylene

Ethylene is a natural plant hormone which plays an important role in the ripening. The ability to produce it varies greatly among fruits; with fruits generating moderate to very high quantities being classified as climacteric (apples, pears, peaches, bananas, mangoes etc.). The production of ethylene is always accompanied by increase in respiration rate. Consequently, high respiration rate stimulates production of ethylene which induces early ripening of fruits leading to their reduced shelf-life.

Oxygen and Carbon Dioxide Concentration

Reduction of O_2 concentration to less than 10 per cent provides a tool for controlling the respiration rate thus slowing down senescence. However adequate O_2 must be available to maintain aerobic respiration. Elevated levels of CO_2 can inhibit ethylene action delaying fruit ripening. At concentrations above 20 per cent, a significant increase in anaerobic respiration occurs which can damage the plant tissue.

Transpiration

All fruits and vegetables continue to loose water through transpiration even after harvesting and this loss of water is the main activity which affects their commercial and physiological deterioration. Transpiration causes wilting, shrinkage, loss of firmness, crispiness and leads to succulence. In other words its fresh appearance, texture and flavour is lost. If humidity is lower than 70 per cent, the loss of moisture due to transpiration becomes faster than loss of solids by respiration rate (Heinze 1961). Water loss of 3-7 per cent is generally enough to cause a noticeable loss of quality and value (Table 9.2). The various factors affecting transpiration are surface area to volume ratio, nature of surface coating, relative humidity, atmospheric temperature and pressure, extent of any mechanical injury/damage etc. As discussed earlier, respiration generates heat, which raises the temperature of the tissue. So water in the tissue is vaporised and therefore transpiration increases. Thus to minimise transpiration, produce should be kept at low temperature.

Table 9.2: Maximum Storage Life and Maximum Percentage Water Loss for Fruits/Vegetables before Becoming Un-saleable

Sl.No.	*Commodity*	*Maximum Storage Life at Optimum Conditions, Weeks*	*Maximum Water Loss Before Becoming Un-saleable, per cent*
1.	Grapes (Thompson seedless)	4-10	2
2.	Spinach	1.5-2	3
3.	Lettuce	1.5-3	3
4.	Cucumber	1.5-2	5
5.	Peas (in pod)	2-5	5
6.	Peaches	2-5	5
7.	Beets with tops	2-4	5
8.	Beans (runner)	1-1.5	5
9.	Blackberries	1.5	6
10.	Peppers	1.5-2	7
11.	Beets/topped	30-40	7
12.	Strawberries	2-3	7
13.	Tomato (25 per cent ripe)	2-3	7
14.	Cabbage (primo)	6-12	7
15.	Carrots/topped	40-50	8
16.	Asparagus	2-3	8
17.	Brussels sprout	4-6	8
18.	Onion	30-35	10

Source: Thompson, A.K. 1996; Hardenburg *et al.*, 1984.

Microbial Decay

Micro-organisms like fungus or bacteria can proliferate and secrete enzymes to bring about quality deterioration. However mechanical damage like bruising is the potential source of microbial activity. High temperatures and high humidity favour the development of this postharvest decay. Produce also stores and absorbs heat from its environment, therefore, proper temperature management is essential to preserve the quality and prolong the shelf-life of fresh produce.

Effect of Delay in Cooling on the Quality of Fruits and Vegetables

Quality loss after harvest occurs as a result of physiological and biological processes, the rates of which are influenced primarily by product-temperature. The product after harvest should always be cooled as quickly as possible to slow down changes associated with ripening. This gives more time to market the product. A few hours just after harvest are very critical to product quality and shelf-life. Each one-hour delay between harvest and cooling results in the loss of one day of shelf-life (Cantwell 1998, Sullivan *et al.*, 1996). Much more deterioration occurs during the first hour after harvest at 25°C than during a week at 1°C (Anonymous, 1981). Perishable products should be brought to their final storage temperature within 24 hours (Anonymous 1967). These underlining facts highlight the necessity of pre-cooling. For example, strawberries experience increasing deterioration losses as delay between harvest and cooling exceeds one hour. With 4 hours delay in cooling from 30°C, about 70 per cent of strawberry crop is marketable; while with 8 hours delay, only 40 per cent of the crop is marketable (Mitchell, 1972). Exposure to high temperature and direct sunlight may cause sunburns and sunscald injury. The temperature above 32°C may cause ripening disorders in commodity like tomatoes (Thompson *et al.*, 2001). Tomatoes exposed to sun for one hour after harvest are 7-15°C hotter than fruits kept under shade (Anonymous 1986).

Shrivelling and loss of fresh glossy appearance are the two most noticeable effects of cooling-delays particularly for commodities that loose water quickly and show visible symptoms at low levels of water loss *e.g.* leafy green vegetables and carrots often loose crispness; broccoli and peppers loose firmness and green onions loose texture.

Cantwell and Thangaiah (2001) reported reduced visual quality, glossiness and increased symptoms of dehydration in bell peppers and eggplants due to cooling delays of more than 9 and 6 hours at 20-25°C respectively. Thompson seedless grapes show visible symptoms of stem shrivelling at 3 per cent water loss or less. Thus the moisture loss in the field should be confined to less than 1 per cent to hide shrivelling. Cooling reduces wilting due to reduced evapo-transpiration at low temperature, *e.g.* at temperature above 32°C, 1 per cent loss can occur in leafy vegetables in about 1 hour but at 21°C cooling can be delayed upto 12 hours. Spraying water on water tolerant products also slows down moisture loss (Thompson *et al.*, 2001).

Cooling-delay tends to increase decay. Brash (1996) reported increased development of rots in carrots due to delayed cooling. Applying decay control treatments immediately after harvest can minimize damage. For example, grapes should be fumigated with SO_2 within 12 hours after harvest in a separate chamber from pre-cooling. Similarly orange and lemons should be treated for decay within 24 hours. Alternatively, grapes can be field-packed in plastic liners to slowdown water loss and a sulphur dioxide generator pad to control Botrytis decay (Thompson *et al.*, 2001). Pre-cooling slows down the deterioration and rotting process by retarding the growth of decay organisms. Table 9.3 may be used as a guide for acceptable time of cooling-delays.

Cooling-delay is not always detrimental but sometimes can be indirectly beneficial for delaying scald development in apples and reduced cracking in carrots. Commodities like peaches, plums, pears and tomatoes are intentionally kept at 16-24°C after packing to promote ripening. If kept at a relative humidity above 85 per cent, these commodities will actually improve in quality in a controlled ripening process. But they must be cooled after ripening has started to reduce mechanical damage during transportation. Similarly, water loss is not always detrimental. Products like carrots need loss of some turgidity to reduce mechanical damage in handling and fresh-cut processing. In green onions, slight water loss reduces curvature problems. Cooling delays for 16 hours leads to proper curing and drying in onions and garlic and wound healing in potatoes (Thompson *et al.*, 2001).

Table 9.3: Recommended Cooling-delay for Fruits and Vegetables

Product	*Allowable Cooling-Delay (hr)*	*Disadvantage of Cooling-delay*	*Advantage of Cooling-delay*	*Comments*
Mushroom	4	Water loss, decay	None	
Spinach	4	Water loss	None	Can be re-hydrated
Leafy Green vegetables	4-8	Loss of water and crispiness	Product is subject to and mechanical injury if cold and turgid	Re-absorb water during hydro-cooling
Grapes	4-8 at < 29°C	Shrivelling, stem browning and increased decay	None	Treat with SO_2 within 12 hr
Cauliflower	8	Water loss	None	
Carrot	8	Loss of water and crispiness	Reduced cracking if carrots are cold and turgid	Re-absorb water during hydro-cooling
Cucumber	8	Water and chlorophyll loss	None	
Green Onion	8	Texture loss	Water loss during cooling-delay may reduce curvature	
Peach	8	Water loss, decay and rapid softening in late season lots	Fruit conditioned at 20°C is less subject to internal browning	
Watermelon	8	Loss of sugar and texture above 27°C	None	
Dry garlic	16	None	Curing and drying	

Contd...

Table 9.3–Contd...

Product	*Allowable Cooling-Delay (hr)*	*Disadvantage of Cooling-delay*	*Advantage of Cooling-delay*	*Comments*
Dry onion	16	None	Curing and drying	
Potato	16	Weight loss if low relative humidity	Curing/wound healing	
Peppers	16	Water loss; loss of firmness, increased decay	Reduced mechanical injury if peppers are cold and turgid	
Tomato	16	Increased decay; color development	Color development	Apply fungicide within 8 hrs, higher pulp temperature can cause decay from dump tank water absorption
Musk melon	16	Pulp softening and premature ripening	None	
Orange	16	Increased rind defects and decay	None	Treat for decay within 24 hr.
Pear	16	Shrivelling near stem during long term storage	Pre-conditioning and ripening at room temperature	
Apple	24	Texture loss and shorter storage life	24–48 hr cooling-delay delays scald evelopment	
Lemon	24	Increased decay	None	Treat for decay within 24 hr.

Source: Thompson *et al.*, 2001.

Why the Produce Needs Cooling

1. To reduce respiratory activity and degradation by enzymes
2. To reduce internal water loss and wilting
3. To slow down or inhibit the growth of decay-producing micro-organisms
4. To reduce the production of natural ripening agent
5. To provide market flexibility by allowing the grower to sell produce at the appropriate time.

Pre-cooling

Pre-cooling is a word coined by Powell and his co-workers in the US Department of Agriculture in 1904 (Ryall and Pentzer 1982, Powell 1905). Since then various definitions have been given but the most appropriate is by Janick (1986) which states that pre-cooling refers to the rapid removal of field-heat from freshly harvested commodities before shipment, storage or processing. The temperature difference between freshly harvested crop and its optimum storage temperature is an indicator of the field-heat. Produce is usually pre-cooled to 7/8 or 88 per cent of the temperature difference. Additional cooling is limited by the time and energy required to reduce the produce temperature to the optimum storage temperature. Most of the storage rooms designed for holding produce under refrigerated conditions have neither refrigeration capacity nor the air movement needed for rapid cooling. Thus pre-cooling is generally a separate operation that requires special equipment and/or rooms. Fordham and Biggis (1985) stated that conventional cold stores are unsuitable for this operation because 75 per cent of the refrigeration capacity may be required simply to remove field-heat. However, Boyette *et al.* (1994) reported that field heat accounts for 58.9 per cent of the total heat load. The cooling rates in cold storages are not better than 0.5°C/hr due to varied quantities of produce stored (Gibbon, 1972). In each of the cooling method, from 20 minutes or less to 24 hours or more may be required for adequate cooling (Bartlett, 1972).

Pre cooling is accomplished commercially by the following methods with many variations and alterations within these techniques:

1. Room cooling

2. Forced air cooling
3. Hydro-cooling
4. Package icing
5. Vacuum cooling and
6. Cryogenic Cooling.

In general, most of the cooling is done at the packing houses or in under the central cooling facilities. While some of the products may be pre-cooled by any of these methods without suffering any loss in quality; others may be affected adversely depending upon the type of cooling technique applied (Wang 1993). For example, cantaloupe may be successfully pre-cooled by several methods but not by vacuum cooling. Iceberg lettuce on the other hand, could be effectively cooled only by vacuum cooling (Hardenberg *et al.*, 1990). The choice of the method depends upon several factors listed below:

Type of Produce

Different types of commodities have different cooling requirements. For example, strawberries, apples and broccoli require near freezing temperatures, whereas such low temperatures would damage summer squash, mangoes, bananas or tomatoes. Ryall and Pentzer (1982) reported that for produce susceptible to chilling injury, cooling to only 10-13°C would be adequate. Similarly because of the problems that can be caused by wetting of certain products, like cut flowers and melons hydro-cooling or icing may not be appropriate (Nowak and Rudnicki 1990). If small amounts of different produce with different cooling requirements must be cooled and stored together, the temperature will have to be set high enough to prevent chill injury of susceptible crop (Table 9.4). Some fruits and vegetables produce ethylene gas as a natural product of ripening and respond to this gas by accelerating their ripening while others are sensitive to it. So it is very important not to cool/store such commodities together (Table 9.5). In addition some products generate odours that are readily absorbed by the other items. The odour of apples, guava and onions are easily transferred to other produce items.

Timeliness in Cooling

Some cooling techniques are faster than the others and therefore have different thorough put of produce. If larger volumes of produce are to be cooled per day, it may be necessary to use faster cooling method. For example, half cooling-time required for apples packed in 18 kg boxes would be 0.33 hour in hydro-cooling compared to 12 hours under room-cooling (Table 9.6). The capacity of hydro-cooler is at least twice as high as that of an air cooler.

Table 9.4: Horticultural Produce with Different Cooling Requirements

Chill Sensitive, Below 40-45°F	*Freeze Sensitive, Below 32°C*
Mangoes	Apples
Beans (all types)	Asparagus
Eggplant	Cabbage
Okra	Peaches
Peppers	Sweet corn
Potatoes	Strawberries
Pumpkins	
Sweet potatoes	
Tomatoes	
Watermelons	

Source: Boyette *et al.*, 1991.

Table 9.5: Horticultural Produce with Different Ethylene and Odour Sensitivity

Ethylene Producer	*Ethylene Sensitive*	*Odour Compatibility Between Commodities*	
		Odour Producer	*Odour Absorber*
Apples	Broccoli	Apples	Cabbage, celery, onions and potatoes
Apricots	Cabbage	Pears	Carrots
Bananas	Carrots	Onions (dry and green)	Grapes, mushrooms, corn
Cantaloupes (*e.g.* Musk-melon)	Cucumbers	Carrots	beans
Kiwis	Cut flowers	Peppers	
Peaches	Egg plant		
Pears	Leafy greens		
Plums	Okra		
Tomatoes	Peas		

Source: Boyette *et al.*, 1991, Gast and Flores 1992.

Table 9.6: Effect of Pre-cooling Method on the Half-cooling Time of Apples Packed Loosely in 18 kg Boxes

Sl.No.	*Cooling Method*	*Half-cooling Time, Hour*
1.	Conventional cool room	12
2.	Tunnel with air velocity of 200-400 m min^{-1}	4
3.	Jet cooling with air velocity of 740 m min^{-1}	0.75
4.	Hydro-cooling (loose fruit)	0.33

Source: Hall 1972.

Product Packaging Requirements

The choice of cooling method is also greatly influenced by the type of packaging used *i.e.*, whether the produce is in the box, bin or bag. The package design and material can have effect on the method and rate of cooling (Stanley 1989). Arifin and Chau (1988) illustrated the effect of vent-hole design on cooling rates of straw berries and found that new design improved the cooling efficiency vastly especially at lower air flow rates. Similarly the area of vents also affects the required cooling time. Ryall and Pentzer (1982) found that venting 5 per cent of the side area of CFB boxes would result in reduction in cooling time by as much as 25 per cent. Sainsbury (1961) reported that packed boxes of certain produce requires several days to reach storage temperature. Consequently the flow rates have to be increased sufficiently (170-225 m^3/min) so as to cool the produce to 5°C in approximately 12 hours (Ryall and Pentzer, 1982).

Efficiency of Cooling Method

Vacuum cooling is more energy efficient and economical besides giving more rapid and uniform cooling but it requires high initial investment and skilled labour. Hydro-coolers can be efficiently operated as vacuum cooler provided heat leakage (≈50 per cent) is taken care of. The package icing and forced air-cooling are the least efficient. Package icing has problems of additional weight of ice used, thereby, adding more weight during transport. In forced air cooling, the fan adds 1/3rd heat in addition to its own energy consumption.

Economics of the Method Used

The construction and operating cost varies greatly among cooling methods. The expenses of the chosen cooling technique must be justified by higher sale prices or other economic benefits (Table 9.7).

Table 9.7: Comparison of Costs of Cooling Systems

System	*Capital Cost*	*Running Cost*	*Through Put (24 hr day or as in bracket)*
Hydro-cooling	100%	100%	30 T
Vacuum cooling			
2 pallet	132%	100%	10 T
5 pallet	180%	100%	25 T
Conventional cold store			
Small	36%	76%	4 (10) T
Medium	64%	76%	8 (25) T
Large	100%	76%	14 (40) T

Source: Ward 2004.

In addition, requirements of skilled labour and availability of desirable utilities are also important. Sophisticated machines require skilled personnel for optimum trouble free operation. An occasional failure of the system can spoil the whole produce. The equipment have to be serviced periodically. Similarly the availability of proper infra structure and utilities like electricity, well metalled road, alternate energy sources and connectivity with production centres will also influence the selection of cooling method.

Rate of Cooling

The rate of cooling of any commodity depends primarily upon four factors, *viz.*, accessibility of the product to the refrigerating medium, difference in temperature between the product and the refrigerating medium, velocity of refrigerating medium and kind of cooling medium. Initially when the product is warm, temperature drops quite rapidly but later on this rate slows down as the temperature of the product decreases. Average product-temperature during the cooling follows a pattern similar to Figure 9.1. In all pre-cooling methods, the term 'half-cooling time' is used. The product is considered to be half-cooled when its temperature drops to half the difference between its initial temperature and the temperature of cooling medium. After another half-cooling period, the product is

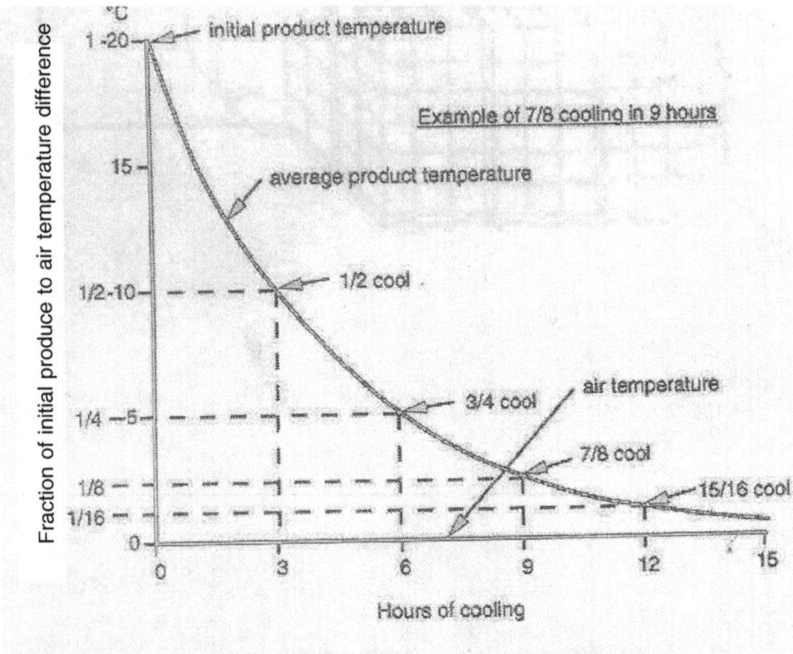

Figure 9.1: Typical Half-cooling Time Curve

three-quarter cooled. Product is usually considered finished for cooling at seven-eighth or fifteen-sixteenth's cooling. The half-cooling time is the same for a given pack exposed in a given method regardless of the initial product and cooling temperature. It is the temperature difference that counts. For example, if a produce pre-cooled in a given method has a half- cooling time of 12 hours, it will cool from 80°F to 60°F [(80-40)/2 = 20°F; 80-20 = 60°F] in 12 hours at 40°F coolant temperature; in the next 12 hours it will cool to 50°F [(60-40)/2 = 10°F; 60-10 = 50°F] and in the next 12 hours to 45°F and so on (Pantastico 1975).

Pre-cooling Methods

Room Cooling

Pre-cooling the produce in a cold store or pre-cooling room is an old but well-established practice (Sainsbury 1961). This method involves placing the warm produce in boxes (wooden, fibre board or plastic), bulk containers or various other packages into an insulated cold room equipped with refrigeration unit. Cooling is achieved by moving the ambient room air around the containers. Containers are stacked individually, arranged in lengthwise channels so that cold air from the ceiling blows over or around the produce to contact all surfaces of the container (Figure 9.2). An airflow rate of approx. 100 cubic feet per minute per ton (5 lit/sec/ton) of storage capacity is necessary to cool the produce. Cooling from outside to the center of stack is mostly by conduction. Once cooling is complete, airflow rate would be dropped to the lowest speed (20-40 cfm/ton) that will keep the produce cool. Prange (1994) stated that air velocity needed for pre-cooling is greater than that needed for storage; consequently products would loose moisture faster than under ideal storage conditions. Therefore desiccation of products can be a problem in room air-cooling, which can be eliminated by providing rooms with humid air (RH>95 per cent). Containers should be well vented (5 per cent) so that air can circulate through them easily. Spacing between the containers and the wall must be 6-12 inches and between the boxes and the ceiling, 18–24 inches. Room cooling is a slow method of cooling and is best suited to less perishable commodities such as potatoes, onions, apples and citrus fruits since perishable crops will not deteriorate too much before being adequately cooled. The main advantage of room cooling is its low cost because no special equipment or facility is required.

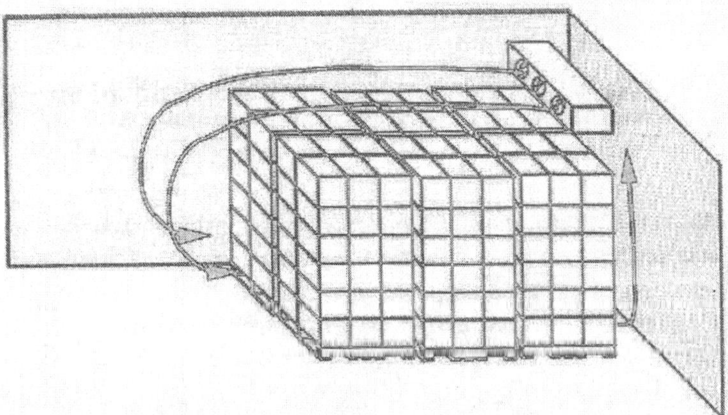

Figure 9.2: Typical Room Cooling System

Low cost rooms can be constructed using concrete floors and polyurethane foam as insulation material. Building a cube-shaped store would reduce the surface area per unit volume of storage space in addition to reduction in construction and refrigeration cost (Kitinoza and Gorny 1998). Effective cooling may require 18-24 hours or even longer (Anonymous 1982). Room cooling is becoming obsolete as more commodities are being handled in large quantities due to better mechanization.

Design of Room Cooling Facility

Although the primary function of any cooling facility is to remove field-heat, however, other important consideration is to provide cold storage space. Although cooling capacity and storage capacity are two separate things but together they determine the size of the room cooling facility. So to estimate the size of the cooling (refrigeration) system, it is important to know the quantity of produce you are going to cool and store. Refrigeration capacity, which is a measure of the rate at which the system will transfer heat, is normally expressed in tons. A ton of refrigeration is the amount of heat required to melt 1 ton of ice in a 24-hour period or conversely a refrigeration system of 1-ton capacity is theoretically capable of freezing 1 ton of water in 24 hours. (1-ton of refrigeration = 288 000 Btu = 12000 Btu/hr.)

The correct size of the refrigeration unit is determined by the following factors:

1. Weight of the produce to be cooled. More the produce to be cooled larger the refrigeration unit should be.
2. Time required from the start to finish of cooling. Cooling the produce faster than required is expensive because the refrigeration system should be larger and the cost of electricity would be more. To cool a load of produce in one hour instead of two hours may require twice the refrigeration capacity and the cost of electricity may be three times higher.
3. Refrigeration space (size), its insulation and how it is operated. Almost half of the refrigeration capacity is used to overcome heat gained through the floor, walls, ceiling and doors. It is important to minimise these losses.

4. Temperature difference between cooling air temperature and field temperature of the produce or temperature drop required.

Storage Capacity

The judgement whether to cool and market the produce immediately or store it for some time often depends not only on the type of produce and market conditions but also on the availability of space in the storage facility. Highly perishable produce requires less storage space than less-perishable commodities as it cannot be held for a long time without losing quality. It is advisable to construct enough storage space for at least one day's maximum harvest of the most perishable commodity. Also it should be kept in mind that it is much easier to build adequate storage space initially than to add it later. Cost of construction decreases and the efficiency in energy utilisation increases with the size of the facility. However, the excess storage space is a wastage of energy and money. The following formula may be used to estimate the size of the refrigerated space to build:

$$V = 2.5 \times (C + S)$$

where,

V = Volume of the refrigerated space (ft^3)

C = Maximum number of bushels (specific volume of the product) to be cooled at any time

S = Maximum number of bushels to be stored at any time.

After estimating V, it is divided by the height of ceiling (ft.) to obtain the cooling room floor-area(ft^2). The ceiling-height should not be more than 18 inches greater than the maximum stack height of the produce. For small to medium cooling rooms, 25 per cent of total floor space is to be reserved for aisles and walkways. Since produce containers should never be allowed to touch the walls, at least an additional 6-inch of space for good air-circulation may be reserved.

Insulation

From the knowledge of thermodynamics we know that heat flows from objects at high temperature to the others at lower temperature till equilibrium is established. However all materials whether good or bad conductor of heat, offer resistance to the flow of heat. Materials that offer high resistance to the flow of thermal energy are called insulators. Selection of proper insulation is one of the most important decisions one has to make. The characteristic of insulators vary considerably and suitability for a given application should be the deciding factor rather than the cost. The most important characteristic of the insulating material is its R-Value which is the measure of the resistance to the movement of heat. Higher the R-Value, higher will be its resistance and better is the insulating property of the material. The R-Value for the common building materials is given in Table 9.8.

Table 9.8: R-Value for the common building materials

Material	*R-Value/ Inch Thick*	*Material*	*R-Value/ Inch Thick*
Wool or fibre glass	3.5	½ inch plywood	1.25
Cellulose	3.5	Plain expanded extruded polystyrene	5.00
Glass wool	2.5–3.0	Expanded rubber	4.55
Wood shavings or saw dust	2.22	Expanded polystyrene	3.57
Cane fibre board	2.50	Aged expanded Polystyrene	6.25
Solid concrete	0.08	Sprayed expanded Urethane	6.25
8-inch concrete block	1.11	Urea form-aldehyde	1.2-5.5
8-inch concrete block with vermiculite in core	5.03	Masonite particle board	1.06
3/8 inch plywood	1.25	25/32 inch insulated sheathing	2.06

Source: Boyette *et al.*, 1991.

Calculating the Heat Load

To make sure that the storage room can be kept at the desired temperature, calculate the required refrigeration capacity using the most severe conditions expected during operation. These conditions include the mean maximum outside temperature; the maximum amount of produce cooled each day and the maximum temperature of the produce to be cooled. The total amount of heat that the refrigeration system must remove from the cooling room is called the heat load. The total heat load comprises of:

1. Field heat–heat removed (vital heat) from the produce as it cools to the storage temperature
2. Heat of respiration–heat generated by produce as a natural by-product of its respiration
3. Heat conduction–heat entering through insulated walls, ceilings and floor
4. Service load–heat from lights, equipments, people and warm moist air entering through cracks and open doors.

Example: Two- thousand bushels of pears are loaded into a storage room 25 feet square and 15 feet high (inside dimensions) at a rate of 200 bushels per day. The walls have an insulation value of R-16, the ceiling a value of R-20 and the floor a value of R-11. The cold room will be operated at 32°F during August when the mean maximum temperature is 80°F.

Solution: First of all let us consider the heat load due to conduction. The heat is conducted into the cooling room through the walls, ceilings and floor. Fourier law of conduction explains the amount of heat flowing through these surfaces.

$$Q_{walls}\cdot = KA\frac{dT}{dX}$$

$$\frac{\text{wall area (sq.ft)* Temp.difference}}{\text{Thickness/K}} = \frac{\text{Wall area * Temp.difference}}{\text{R} - \text{Value}}$$

$$Q_{walls} = \frac{4 \times 25 \times 15 \times (80 - 32)}{16} = 4500 \text{ Btu/hr}$$

Similarly the heat conducted through ceiling =

$$\frac{25 \times 25 \times (80 + 10 - 32)}{20} = 1812.5 \text{Btu/hr}$$

(Ceiling receives direct sunshine and its temperature is about 10°F more than the outside)

And heat conducted through floor =

$$\frac{25 \times 25 \times (55-32)}{11} = 1306.8\text{Btu/h}$$

The floor temperature is about 55°F for most of the time.

So the total heat conducted through walls, ceiling and floor

= (4500 + 1812.5 + 1306.8) Btu/hr

= 7619.3 Btu/hr

Field Heat

It is the product of specific heat, the difference of temperature between the field temperature and storage temperature and the weight of the commodity. The specific heat of the product for practical purposes can be estimated from the following equation:

Specific heat = 0.008 × (per cent water in food) + 0.20

Therefore the specific heat of pears with moisture content of 84 per cent will be = 0.008 × 84 +0.2 = 0.87

The specific heat of water is 1 Btu/Ib°F. Since fruits and vegetables contain large proportion of water, their individual specific heat is directly related to its moisture content. Table 9.9 given below lists the actual values of specific heat. A bushel represents specific volume of a commodity and its value is different for different products, *e.g.*, a bushel is 1.25 ft^3 for small grains or 2.5 ft^3 of ear corn. Similarly a bushel of pears (weight of fruit in one plastic crate) weighs about 45 pounds. But for the given problem it can be approximated as 50 pounds to take into the weight of container, which also is cooled.

The field heat becomes = 0.87 × (80-32)* 200*50 = 417600 Btu

Since the pears are to be cooled in one day, therefore hourly field heat = 417600/24 = 17400 Btu/hr

Heat of Respiration

Horticultural commodities give off heat as they respire and the amount of heat produced depends on the temperature, nature of the crop and conditions and treatment the crop receives. The amount of heat given off approximately doubles at 18°F (10°C) rise in temperature. Therefore less refrigeration is required to remove the heat of respiration when produce is cool than when it is warm.

Table 9.9: Specific Heat and Heat of Respiration for the Horticultural Crops

Commodity	*Specific Heat (Btu/lb °F)*	*Respiration (Btu/lb/hr)*	
		Cool	*Warm*
Apples, Summer	0.87	0.018	0.340
Apples, fall	0.87	0.012	0.240
Asparagus	0.94	0.94	2.523
Beans, butter	0.73	0.123	0.716
Beans, string	0.91	0.161	0.885
Beets	0.90	0.028	0.092
Broccoli	0.92	0.092	1.376
Cabbage	0.94	0.023	0.257
Cucumbers	0.97	0.119	0.17
Grapes	0.86	0.014	0.179
Green onions	0.91	0.096	0.704
Leafy green vegetables	0.90	0.100	1.034
Okra	0.92	0.257	1.583
Peaches	0.91	0.023	0.466
Peas, field	0.73	0.160	1.554
Pepper	0.94	0.046	0.252
Potatoes	0.84	0.028	0.055
Strawberries	0.92	0.069	0.872
Sweet corn	0.79	0.186	1.644
Sweet potatoes	0.75	0.064	0.100
Tomatoes, mature	0.94	0.030	0.197
Turnip	0.93	0.034	0.130
Watermelon	0.94	0.034	0.110

Source: Boyette *et al.*, 1991.

On the last day of harvest, the storage will contain 1800 bushels of cool pears and 200 bushels of field warm fruit. Assuming pears to be summer apples, the Table 9.9 indicates that 0.340 Btu/Ib/hr heat is generated whereas cooled fruit produces only 0.018 Btu/Ib/hr. Since the warm fruit will cool in one day, so the average of these values can be taken *i.e.* (0.340 + 0.018)/2 = 0.179 Btu/Ib/hr.

So, heat generated by cooled fruit = 1800 × 50 × 0.018 = 1620 Btu/hr and heat generated by warmed fruit brought from field = 200 × 50 × 0.179 = 1790 Btu/hr.

Total heat of respiration= 1620 + 1790 = 3410 Btu/hr.

Service Load

There are a number of miscellaneous items in the cold room facility and heat generated by these is called service load. It includes heat given off by equipment like lights, fans and the people working there. Heat is also brought in when the door of the cold room is opened. The amount of heat contributed by these individually is very difficult to assess. Service load is therefore collectively approximated as equal to 10 per cent of heat from the other sources.

Service load = 10 per cent (7619.3 +17400 +3410) = 2842.9 Btu/hr

Hence total heat load = 7619.3 + 17400 + 3410 + 2842.9 = 31272.2 Btu/hr

Refrigeration capacity = 31272.2/12000 = 2.60 tons

This means 2.60 ton refrigerated system has to work 24 hours to provide the desired temperature. But if our system is operated 16–20 hours in a day then the refrigeration capacity has to be increased. For example, for 16 hours operation the refrigeration system capacity needs to be raised by 24/16 = 1.5 times to 3.9 tons. Further to compensate chilling rate, a factor of 1.5 is used for most fruits and vegetables. (The rate at which heat is removed is not uniform during the cooling cycle but is greatest at the beginning).

So the actual refrigeration capacity needed for efficient cooling = 3.9 x1.5 = 5.85 tons.

Forced Air Cooling:

Forced air-cooling was developed by Guillou (1960) to cool products requiring relatively rapid removal of field-heat immediately after harvest. It is a modification of room cooling which often does not remove field-heat rapidly in highly perishable crops. This heat is often removed from only near the outside of the container while in the centre it is generated at the faster rate by product respiration causing temperature to rise. In such situations room cooling is not an advisable option. To maintain the quality and freshness, produce should be cooled to optimum storage temperature by forced air-cooling method which is much faster than room cooling.

Advantages of Forced Air-cooling

1. It decreases the time the produce remains at higher temperature thereby reducing deterioration
2. Shorter cooling time results in better utilisation of the cooling facility
3. It can cool bulk boxes or pallet loads efficiently
4. It is more energy-efficient than room cooling especially when large volumes of produce are cooled.
5. Room cooling facility can be easily modified into forced-air-cooling with a meagre investment on fans etc.

Operation of Forced Air-cooling System

Forced air-cooling is similar to room cooling as the produce is placed in a cold storage room. However, the cool air is forced through the produce instead of around the package. This is accomplished by exposing packages of produce to higher air pressure on one side than on the other. This pressure difference forces the cool air through the packages and past the produce. This allows the air to pick up the heat thereby increasing the rate of heat transfer. Depending upon the temperature, airflow rate and type of produce being cooled; forced air-cooling can be 4 to 10 times faster than room cooling. Hydro-cooling and vacuum cooling are usually 2-23 times faster than forced air cooling (Figure 9.3). Stacking patterns, location and size of container vents are the important considerations. Container vents should be aligned whether the containers are column stacked or cross-stacked. Cooling time depends upon the airflow rate, temperature difference between the produce and the cold air, product configuration (bulk or packaged), carton vent area, depth of product cooled and produce diameter (Linday *et al.*, 1983). Small produce such as berries (about 1inch dia.) will cool almost six times faster

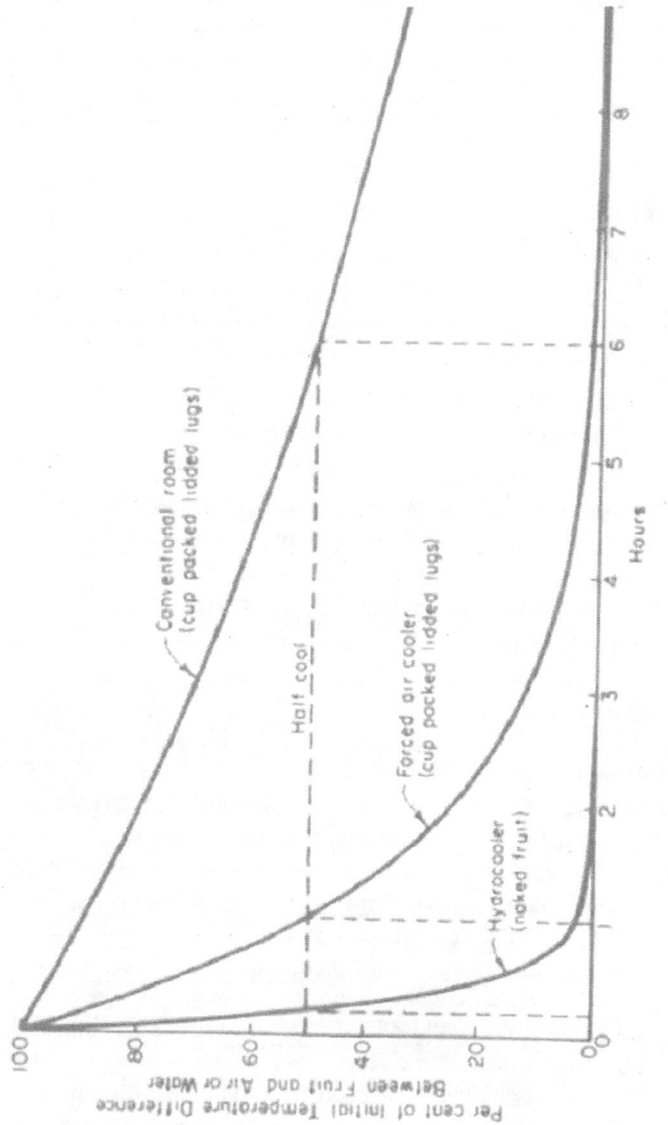

Source: Guillou 196

Figure 9.3: Comparative Cooling by Different Methods (Guillou 1960)

than 6-inch diameter melons (Kitinoja and Gorny 1998). Cooling increases with the increase in air velocity (air flow rate). Figure 9.4 shows the response of a produce to airflow rate. It is clear from the graph that the slope of the cooling curve which represents rate of cooling increases more rapidly at higher air velocity than still air. Since the cooling rate for a forced air-cooling is much greater than for room cooling, a refrigeration system of larger capacity may be required.

Hall (1972) reported that half-cooling time required for apples was 1.25 hours at an air flow rate of 40 m/min as compared to half an hour at 400 m/min. The fan needs to be big enough to provide an optimum airflow rate of 1 cft/m/lb (1 litre/sec/kg) of produce against a wide range of static pressures. Doubling the airflow rate will speed cooling slightly but would need a much larger and costlier fan (6-7 times more hp). Reducing the air flow rate by half will increase the cooling time by about one hour for most commodities but would dramatically reduce fan motor size and energy costs. The operation of forced air cooler should be stopped as soon as produce is cooled. The energy requirement for running the fan could be reduced provided vent area is atleast 4 per cent of the surface area of the cartons. Most fibre-board cartons can be vented up to 5 per cent without significant loss of stacking strength. Non-circular vents 1 cm wider and 4-7 cm should be made away the from corners

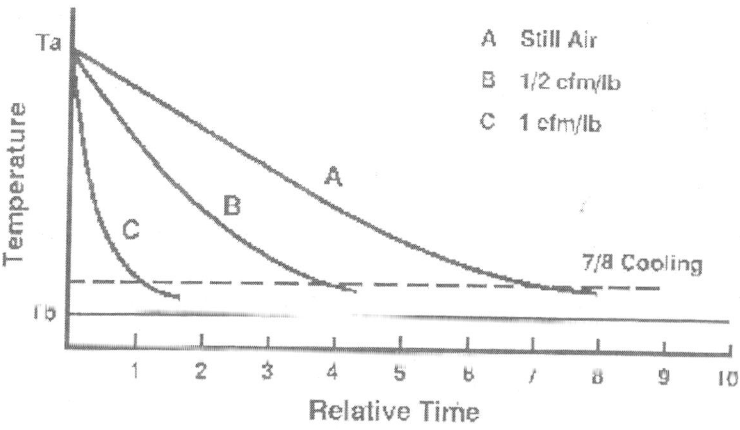

Figure 9.4: Effect of Air Flow Rates on Cooling Time (Boyette *et al.*, 1989)

(Kitinoza and Gorny 1998). The main drawback of forced air-cooling is that it requires definite stacking patterns and high-speed cold air can cause serious water loss. This can be prevented by using humid air (RH>95 per cent). Various methods of humidifying the cooling air have been devised. Thompson and Chen (1988) examined the energy used for a variety of pre-cooling techniques and found that forced air cooling gives the lowest efficiency due to high levels of heat input. Baird *et al.* (1988) reported that when very rapid cooling is required, forced air-cooling is more costly than other pre-cooling methods. Forced air cooing may be used as an alternative cooling method for cooling fruits and vegetables, such as melons and cut flowers that can not be hydro-cooled or are not adaptable to vacuum cooling (Hardenberg *et al.*, 1990).

The followings are the popular forced air variations:

Cooling Wall

In this case a permanent false wall is constructed to form an air plenum. An exhaust fan is fitted in this wall that draws air from the room and directs it over the cooling surface (Figure 9.5). The wall is at the same end of the cold room as the cooling surface. The wall is built with damper system that opens only when containers with openings are placed in front of it. The fan pulls the cold room air through the container thus cooling the produce.

Forced Air Tunnel

In this system, an exhaust fan is placed at the end of the aisle of two rows of containers on pallets. The aisle top and ends are covered with plastic or canvas creating a tunnel. The exhaust fan draws cool room air through container vents and top (Figure 9.6). The exhaust fan may be portable or it may be part of a stationary wall adjacent to the cooling surface.

Serpentine Cooling

Serpentine air system is designed for cooling produce in pallet bins. Stacks of even numbers of bins are placed against a negative pressure plenum wall. Bins have vented bottoms with or without side ventilation. Every other forklift opening, sealed with canvas, matches a cold wall opening. The alternate unsealed forklift openings allow cold air to circulate through the produce (Figure 9.7).

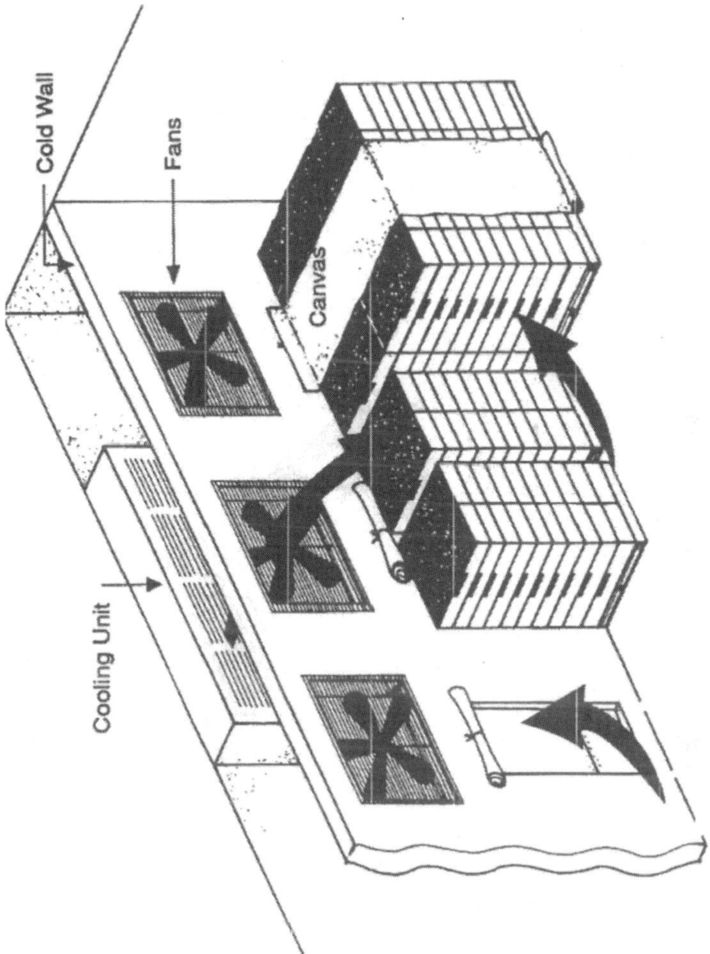

Figure 9.5: Schematic Arrangement of Cooling Wall (Gast and Flores, 1991)

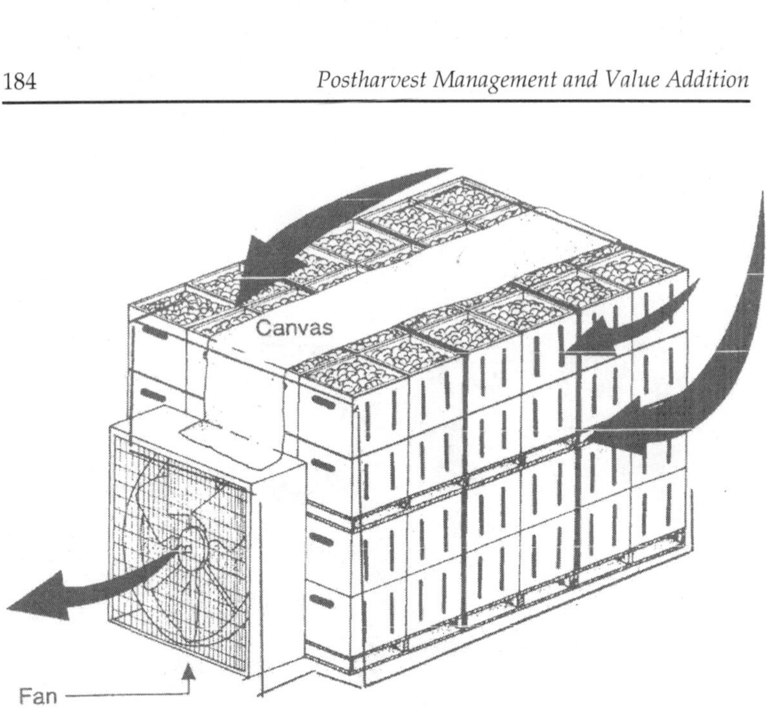

Figure 9.6: Schematic Arrangement of Forced Air Tunnel (Gast and Flores, 1991)

Hydro-cooling

Hydro-cooling has been used since 1923 when it was developed as an outgrowth of celery washing (Showalter and Greison 1972). The process of cooling warm produce directly by chilled/cold water is known as hydro-cooling. It is accomplished by flooding, spraying or immersing in cold water (Athey and Dennis 1991). The transmission of heat from a solid to a liquid is faster than the transmission of heat from a solid to a gas. It is highly popular pre-cooling method because of its simplicity, rapid and effective cooling. Produce items that have large surface area to volume (or mass) and that are difficult to cool can be quickly and effectively hydro-cooled. Unlike air-cooling, no water is removed from the produce. In fact slight wilted produce like green leafy vegetable, may sometimes be revived. At typical flow rates and temperature difference, water removes heat about 15 times faster than air (Boyette *et al.,* 1994). According to Balls (1986) field-heat can be removed in 20-30 minutes using hydro-cooling instead of several hours needed in force air-

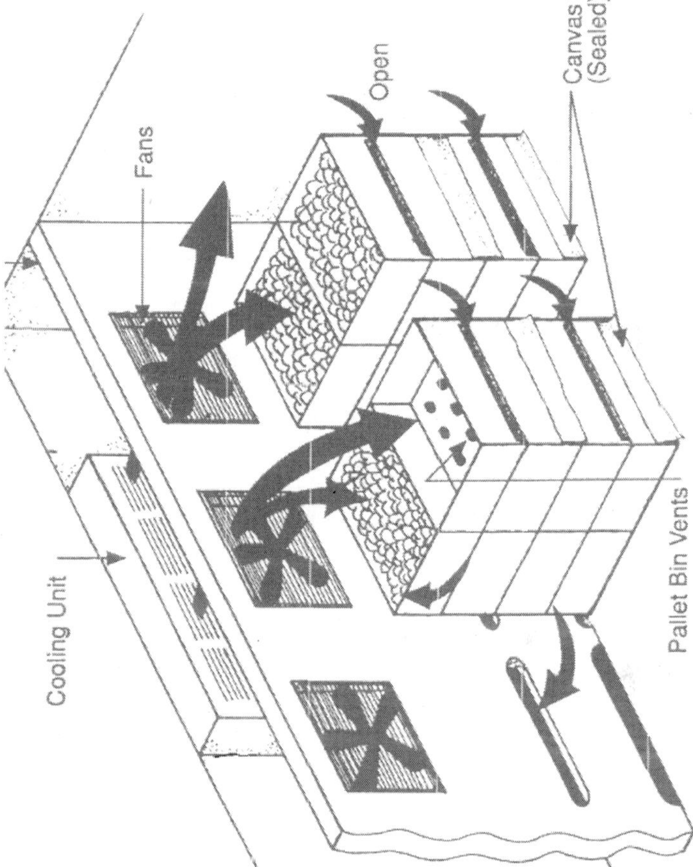

Figure 9.7: Schematic Arrangement of Forced Air Serpentine Cooling (Gast and Flores, 1991)

cooling. High efficiency of the system is achievable due to large heat capacity and high rate of heat transfer of 500 w/m^2°C for agitated water as compared to 5 w/m^2°C for slow moving air (Janick 1986).

Benefits

1. Cools produce rapidly (about 15 times faster than forced air)
2. Allows greater harvesting and marketing flexibility
3. Easily handles large amounts of produce
4. Maintains freshness of produce by preventing water loss

Limitations

1. Can be used only for the produce not sensitive to wetting.
2. Some hydro-coolers are not energy efficient and therefore may not be cost-effective.
3. There are some restrictions on the types of packages and stacking arrangements used for the produce that is to be hydro-cooled.

Operation

In most hydro-coolers, a pump moves the chilled water into intimate contact with warm produce. During this contact, chilled water picks up heat from the warm produce and becomes warmer. The warm water is then again cooled and recycled. For cooling the water, many hydro-coolers have a refrigeration system while others use crushed or chunk ice. Typically large chunks of ice are trucked from ice plants, crushed and added as needed to a warm water reservoir attached to the hydro-cooler. The capital cost of a hydro-cooler of this type is much less than the one with an integral refrigeration system and is preferred by growers with a limited amount of produce or a short cooling season. Mechanical refrigeration is preferred in packing houses cooling more than 375 tons (25 cars) per year

Rate of cooling depends upon the size of the produce, depth of packing, field temperature of the produce, cooling water temperature, flow rate of water and type of container used. For example asparagus spears being long and narrow, can be hydro-cooled in about 2 minutes but capsicum which are large and globular take about

10 minutes to cool (Thompson 1996). Small diameter produce such as radish will typically be cooled in less than 10 minutes; medium sized produce such as peaches, apples etc would take about 20-25 minutes while large diameter produce such as melons take one hour (Kitonoja and Gorny 1998). Water flow rates of 200 $l/min/m^2$ is sufficient if fruits are kept not more than 8" deep and 400 $l/min/m^2$ if depth of packing is 8-15". (Bennett *et al.*, 1965). The refrigeration capacity also depends upon the amount of produce you intend to cool and temperature drop required (Table 9.10). Very little information is available on small scale hydro-coolers; but if you want to cool 100 lb of produce per hour, you will need about 0.4 tons of refrigeration capacity for an 11°C temperature drop; 0.8 tons for 24°C temperature drop or 1.5 ton for a 39°C temperature drop (Thompson 1992). Commercial hydro-coolers can deliver water at a rate of 10-15 gallons per square foot/min (400-600 litre/m^2/min). A lower flow rate will be sufficient if you are cooling a single layer of bulk produce. Insulation of hydro-cooler is necessary, for tests have shown that about 50 per cent of the refrigeration is lost through heat leakage. A hydro-cooler installed in a cold room will require about 30 per cent less refrigeration capacity than un-insulated cooler installed outside (Thompson 1992, Ryall and Pentzer 1982).

Table 9.10: Refrigeration Requirements for Hydro-cooling 2180 kg (100 bu) Peaches for Different Temperature Drops

Sl.No.	*Temperature Drop*		*Heat Load*		*Ice Required*	*Refrigeration Required per Day*
	°C	*°F*	*Kcal/h*	*Btu/h*	*kg/h*	*tonnes*
1	8.3	15	20,866	82,800	261	6.3
2	11.1	20	27,821	110,400	348	8.3
3	13.9	25	34,776	138,000	435	10.4
4	16.7	30	41,731	165,600	522	12.5
5	19.4	35	48,686	193,200	609	14.6
6	22.2	40	55,642	220,800	696	16.7
7	25.0	45	65,597	248,400	783	18.8
8	27.8	50	69,552	276,000	870	20.9
9	30.5	55	76,507	303,600	957	22.9
10	33.3	60	83,462	331,200	1044	25.0

Source: Bennett *et al.*, 1965.

Hydro-cooling can only be used for water tolerant products and those for which washing is a part of their market preparation–mango, radish, cherry, sweet corn, celery, asparagus, green pea, tomato, leafy vegetables and peach (Kitinoja and Gorny 1998, Ryall and Pentzer 1982,Thompson 1996). Generally wetting fruits promotes decay development. Chiefly for this reason berries, grapes, cherries, plums, melons and cut flowers are subjected to spoilage. Capsicums, which were hydro-cooled and even chlorinated, had a higher incidence of rotting during subsequent storage than those which were not (Hughes *et al.*, 1981). Fruits cooled with water are apparently more susceptible to decay upon warming. Some times chilling injury may occur in oranges after pre-cooling (Eaks 1956). In mandarin varieties of citrus there is damage due to water soaking. Another difficulty in hydro-cooling produce is the damage water does to the containers and packing material. Pre-cooling before packing would solve this problem. But loss of refrigeration during pack house operation would be disadvantageous. Hydro-coolers are excellent sources of contamination, therefore water should be changed daily and treated with fungicide and chlorine to prevent spread of decay. Water should be chlorinated with 100-150 ppm of available chlorine (60 ml/40 litre or 2 oz of chlorine bleach/10 gallons of water). Moreover, organic matter ties up with chlorine thus requiring more chlorine than usual. Water can be treated with sodium-o-phenylphenate and newer fungicides benomyle and benlate for controlling decay (Van Blaricom, 1959, Smith, 1957, Ryall and Pentzer 1982, Thompson *et al.*, 1998).

Types of Hydro-coolers

There are various hydro-cooler designs and these differ in their cooling rates and overall process efficiency. Two basic systems of hydro-cooling are commonly used *i.e.* batch and conveyor.

Batch Type Hydro-coolers

These hydro-coolers do not have conveyors. Palletised cartons or bulk bins of produce are stacked into the chamber with a fork lifter. The door of the chamber is then closed and large quantity of chilled water is flooded over the top of the produce, collected at the bottom and again cooled and recycled. Most batch type hydro-coolers can cool only one pallet of produce at a time. These hydro-coolers have a smaller capacity than conventional hydro-coolers and therefore may be less expensive. These are suited to growers with a limited amount of produce that could not economically justify a larger unit (Figure 9.8).

A portable batch hydro-cooler called truck hydro-cooler is widely used in many places. In this method water is cooled and delivered to the produce by a simple system consisting of a tank holding several thousand gallons of water, a high capacity pump, associated piping and valves and an ice crusher. Graded and packaged produce is loaded from the field into the tractor-trailer. Perforated pipes are then inserted above the load to deliver a shower of chilled water from the stationary cooling system. Water coming

Figure 9.8: Batch Type Hydro-cooling (Thompson 1998)

out of the trailer is collected, recorded and recycled. Water flow rate of 1000 gallons per minute for this type of hydro-cooler are common. After an hour of hydro-cooling the pipes are withdrawn, a layer of crushed ice is added over the top of crates and produce is immediately shipped.

Conveyor Type Hydro-cooler

The conventional conveyor hydro-cooler type allows the produce (in bulk or in cartons) to pass along a conveyor under a shower of chilled water. Warm produce is placed on one end of the conveyor and cooled produce is removed at the opposite end. The rate at which the produce is moved through the shower is about 1 foot per minute (Figure 9.9). Manufacturers of conventional hydro-coolers often specify their different models by length. However specifying the capacity of a hydro-cooler entirely by length can be misleading if conveyor speed, conveyor width, water temperature and flow rate are not taken into consideration. Hydro-cooling requires large quantities of water to be passed through the produce. Water flow rates as great as 20 gallons per minute per square foot of active cooling area are common.

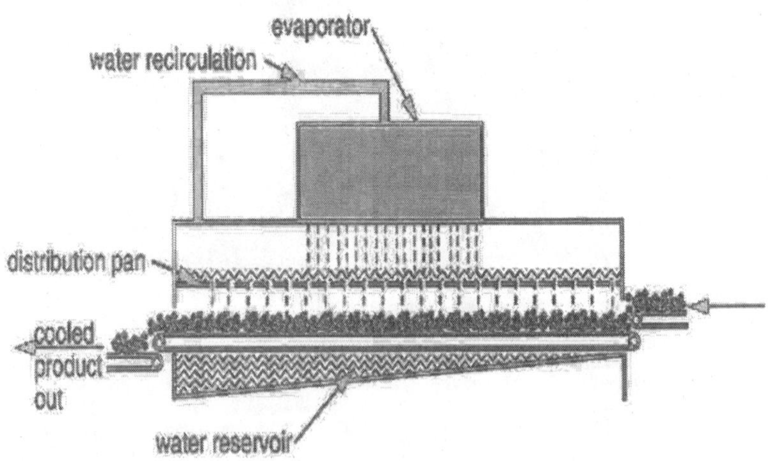

Figure 9.9: Conventional Hydro-cooling (Thompson 1998)

The modified version of conveyor type hydro-cooler also known as immersion hydro-cooler uses a combination of immersion and flood cooling. Immersion hydro-coolers are large, shallow, rectangular tanks that hold the moving chilled water. Crates or boxes of warm produce are loaded into one end of the tank and moved by a submerged conveyor to the other end where they are removed. Crushed ice or refrigeration system keeps the water cold and a pump keeps the water in motion. The length of time the produce remains in water varies with initial conditions and the desired final temperature (Figure 9.10). Grizzell and Bennett (1966) reported that immersion hydro-coolers are nearly twice as rapid as conventional ones due to the fact that moving chilled water facilitates quicker temperature reduction. Boyette (1994) reported that immersion hydrocooler allows greater packaging flexibility *i.e.*, packing after cooling and therefore, it may be more suitable to wide range of growers.

Package Icing

Ice has been used since early times to remove heat rapidly from food and to extend its shelf-life. It is a preferred method of cooling for many types of extremely perishable produce. It is commonly applied in boxes containing produce. It can be used only with water tolerant, non-chilling sensitive products and with water tolerant packages (laminated or waxed impregnated CFB boxes, plastic or

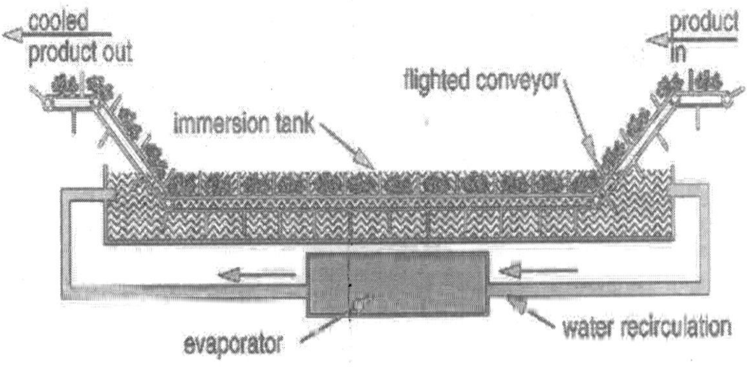

Figure 9.10: Batch Type Hydro-cooling (Thompson 1998)

wooden). The main benefit of package icing is that it not only removes field-heat and maintains low product temperature but also maintains freshness in transit. Roughly one kg of ice would drop the temperature of produce 3 times its weight by about 28°C (Kitinoza and Gorny 1998). Growers with small and large operations can use crushed and liquid ice cooling methods effectively. Block ice must be crushed before it can be used for icing. Pieces of ice should be no larger than 3/8 inch so these can will penetrate the voids in the produce package. Also small pieces are much less likely to cause mechanical damage to the produce. Electrically powered ice crushers can crush large blocks of ice to a suitable fineness. Package icing may be used as an alternate to refrigerated transport for short duration journey. Thompson et a l (1998) consider icing as an energy inefficient and expensive cooling method.

Icing Methods

Crushed Ice Packaging

The simplest icing method is to add a measured amount of crushed ice manually to the top of the carton filled with the produce. This method is sufficient for cooling but in many cases it can result in uneven cooling because the ice generally remains where it was placed until it has melted. In the improved method crushed ice is placed in between the layer of the produce and at the top. Since each carton must be opened, iced and closed again, the process is slow and labour intensive. It is not considered for high volume production. This method is effective for cooling products that are not harmed by contact with ice. Spinach, broccoli, brussels, sprouts radish, green onions, carrots and cantaloupe are the products commonly marketed with crushed ice in shipping containers. To cool the produce from 35°C to 2°C requires the melting of ice equal to 38 per cent of the product weight. Additional ice must be added to account for heat leakage into the package from outside (Hardenberg *et al.*, 1990). For example, atleast 10 lbs of ice must be added to a 20 lb carton of broccoli in order to provide cooling during long distance transport (Kitinoza and Gorny 1998). Similarly, a packed crate of 30 kg lettuce would need 15 kg of ice (Janick 1986).

Liquid Icing

It is also known as slush icing and is a preferred method when large amounts of produce are to be iced in a relatively short period of time. The required equipment consists of an ice crusher, a slurry tank with mixer, a pump and delivery hoses. A typical slurry is made by mixing 40 per cent water and 60 per cent finely crushed ice(1:1.5) plus 0.1 per cent sodium chloride to lower the melting point of ice to affect faster cooling. Water ice ratio may vary from 1:1 to 1:4 (Thompson, 1996). This method may be considered as combination of package icing and hydro-cooling. In the simplest form of liquid icing, a mixture of water and finely crushed ice is pumped into open crusher travelling along a conveyor under an injection nozzle as shown in Figure 9.11. This method is a significant improvement over hand icing because the water carries the ice into the voids throughout the package, thus bringing it into greater contact with the produce. With proper equipment, two workers can apply liquid ice to a pallet of 30 cartons in 5 minutes. The equilibrium temperature of a mixture of melting ice and water is 32°F (0°C). Liquid icing is an excellent method, although it does wet the produce. The surface of the warm, wet produce provides an excellent site for postharvest diseases to develop. Therefore it is essential that once iced, produce should not be allowed to re-warm. Use of chlorine as a disinfectant at a rate of 55-70 ppm is recommended

Top Icing

Here ice is placed on the top of the packed container. Top icing does not do much cooling but it is used primarily as a supplement to principal pre-cooling methods (hydro-cooling, vacuum cooling, forced air cooling) for crated sweet corn, celery and leafy vegetables and for radish and carrots pre-packed in film bags (Hardenberg *et al.*, 1990). The use of top icing has decreased because CFB boxes have replaced wooden or plastic crates. Wax impregnated CFB cartons have allowed the use of icing after packing of products. However, it is being replaced by hydro-cooling or vacuum cooling. Gillies and Toivonen (1995) reported equal water loss from broccoli by hydro-cooling and top icing.

Vacuum Cooling

Vacuum-cooling is achieved by evaporation of moisture from the produce surface. The evaporation is encouraged by reducing the pressure to a point where boiling of water takes place at low temperature. The underlining principle of vacuum cooling process is that water boils at 100°C at atmospheric pressure. With the

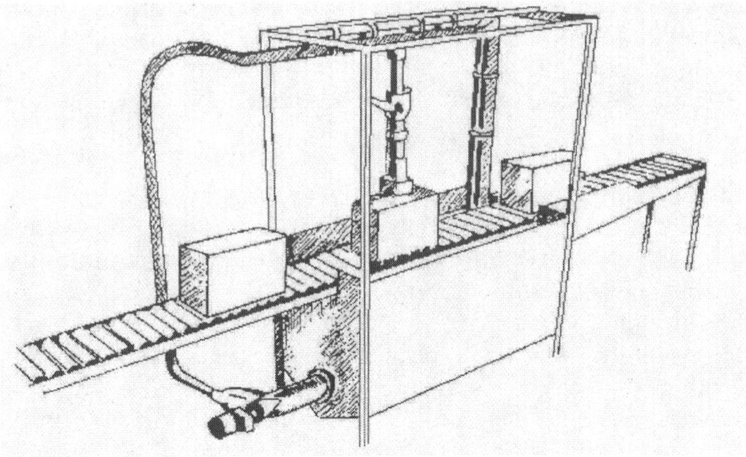

Figure 9.11(a): Liquid Icing in Individual Cartons
(Boyette *et al.*, 1992)

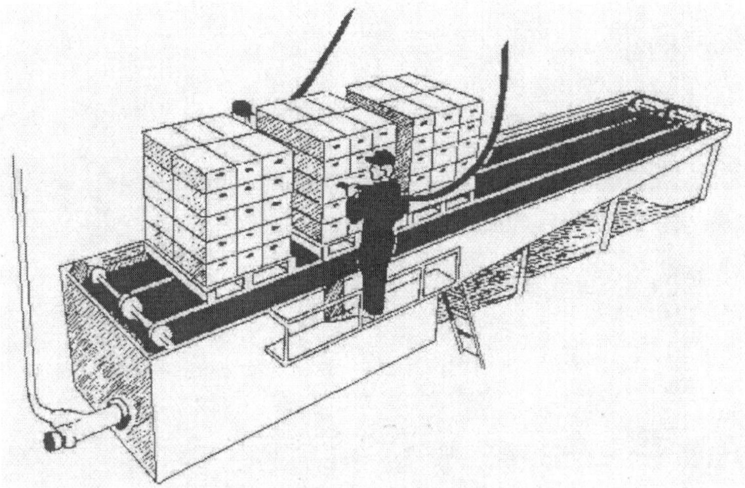

Figure 9.11(b): Liquid Icing in Pallets
(Boyette *et al.*, 1992)

reduction in pressure, boiling temperature reduces and at a pressure of 4.6 mm of mercury, water will boil at 0°C (Anonymous 1982, Barger 1961, Friedman and Radspinner 1956).

Operation

The produce is enclosed in an airtight chamber and low pressure is created by the vacuum pump. Vacuum-cooling process itself occurs in two distinct phases (Love-lidge 1972, Robertson 1978). In phase I, pressure in vacuum chamber is reduced from atmospheric to about 20 mbar. During this phase, evaporation is slow and relatively little cooling takes place *i.e.*, temperature of the produce remains constant. At approximately this pressure called flash point, the produce begins to loose moisture and cool rapidly. In the second phase, the pressure is reduced continuously until pressure corresponding to the desired final temperature is reached (Anonymous. 1998, DiRisio 1994). In practice, most operators do not reduce pressure below 6.09 mbar (saturation pressure corresponding to a temperature of 0°C because of freezing potential at reduced pressure and extra work involved (Zemansky *et al.*, 1975).

Vacuum-coolers are rugged in construction. These are made of heavy duty steel and usually cylindrical in shape to withstand low pressures. Vacuum chambers vary greatly in size; small portable units have a capacity of few pellets while large stationary ones can hold up to 2 carloads. The moisture loss during vacuum-cooling of vegetables ranges from 1.5 to 5 per cent. Since this moisture loss occurs equally from all parts of the produce being cooled, wilting is not apparent unless water loss exceeds about 5 per cent. To avoid it, products are pre-wetted before vacuum-cooling. A special vacuum-cooler called hydrovac has been developed which has a built-in water spray system activated towards the end of cooling and thus minimizes the weight loss (Thompson 1996). About 1 per cent of the weight is lost for every 5-6°C (10°F) lowering in temperature (Barger 1962, Robertson 1978).

The rate of cooling and final temperature attained during vacuum-cooling is affected largely by surface area to mass (or volume) ratio, the ease with which product gives up water from tissues (Barger 1961, Thompson 1996). Thus, fruits and vegetables differ widely in their adaptability to vacuum-cooling. Vacuum-cooling is particularly suitable for leafy vegetables like spinach, lettuce, green onion and carrot tops due to their large surface area to volume ratio (SV). Some products like asparagus, broccoli, cauliflower, celery and sweet corn having medium SV ratio can be pre-cooled if pre-wetted or misted to supply some water needed for vacuum-cooling. Vacuum-cooling is not recommended for products like topped carrots, potato, tomato, cucumber having lower SV ratio (Table 9.11) and are better adapted to other cooling methods. Large temperature gradients develop in products with low SV ratio. To prevent surface freezing before the product is cooled to the required temperature, bouncing procedures is followed by switching the vacuum off-and-on to keep the saturation temperature above freezing (Anonymous. 1994). Crop like tomato, which has a relatively thick wax cuticle is not suitable for vacuum-cooling (Thompson 1996). Vacuum-cooling gives more rapid and uniform cooling compared to other methods of cooling. Also it is more energy efficient and economical as compared to other methods. Turk and Celik (1994) found that lettuce can be cooled in 20-29 minutes in vacuum-coolers compared to several hours in conventional cooling rooms. Kim *et al.* (1995) reported that vacuum-cooling followed by refrigeration at 0°C gives a shelf-life of 40 days compared to 20 days at 0°C for lettuce marketed by conventional methods. Vacuum-cooling can cool mushroom in about 20 minutes compared to 8-13 hours required to bring temperature to 1°C in a cold store (Atkin 1974). Strawberries can be cooled in 22-26 minutes even if they are packed prior to cooling (Pasquatone 1986). For an economical and efficient facility, vacuum-cooling requires large quantities of produce. Buffington *et al.* (1981) described an alternate use of vacuum-cooler for curing of onions and garlic. This extends the cooler use and helps to distribute the cost, therefore, making the technique more economically viable.

Cryogenic Cooling

The use of latent heat of evaporation of liquids nitrogen or solid CO_2 can produce boiling temperature of–196 and–78°C respectively. This is the basis of cryogen pre-cooling. In cryogenic-cooling, produce is cooled by conveying through a tunnel in which liquid nitrogen/ CO_2 evaporates. Since the produce may freeze at this temperature so evaporation rate and conveyor speed are carefully controlled. Cryogen cooling is relatively cheap to install but expensive to run. It is relatively used in soft products like cut flowers, which have seasonal production. The high cost of liquid nitrogen, dry ice and other suitable non-toxic refrigerants makes this process suitable for relatively expensive products (Brosanan and Sun 2001).

Table 9.11: Effectiveness of Vacuum-cooling Vruits and Vegetables with Different Surface-surface Area-to-volume (SV) Ratios

Sl.No.	*Surface Area to Volume Ratio and Commodity*	*Container Type*	*Duration of Vacuum (min)*	*Commodity Temperature*			
				Initial		*Final*	
				°C	*°F*	*°C*	*°F*
1	Large SV ratio						
	Brussels sprouts	Quart cup	20	20	68	3	38
	Endive	Crate	20	20	68	2	36
	Lettuce, head	Carton	13	22	71	2	36
	Spinach	Basket	10	19	67	3	37
2	Medium SV ratio						
	Beans, snap	Hamper	20	27	80	16	60
	Cauliflower	Crate	20	24	76	7	44
	Celery	Crate	13	21	70	8	47
	Sweet corn	Crate	20	28	83	6	43
3	Small SV ratio						
	Carrots, topped	Crate	45	19	66	16	60
	Cucumbers	Basket	20	26	78	23	73
	Potatoes	None	30	18	65	14	57
	Tomatoes	Cello-tray	20	25	77	22	71

Source: Friedman and Radspinner, 1956.

All these techniques described above are used where refrigeration is available in abundance. Areas where refrigeration is either not available or costly, alternative-cooling strategies like shading of produce, night air-ventilation, radiant cooling, evaporative cooling, use of well water, high altitude storage etc. can be quite useful.

Energy Use in Cooling Operations

Cold chains are essential component of horticultural postharvest infrastructure. It ensures maintenance of freshness of produce for extended period of storage. But it has to be seen through the prism of economic viability. The energy is the major component in cost calculation of cold chain operations. Energy conservation regimes can help in cost cutting. Energy use in pre-cooling facility is affected by the amount of heat the refrigeration equipment must remove and the efficiency of the equipment. The main sources of heat are: field-heat and respiration of stored product (vital heat), heat transmission into store through roof and walls, evaporator coil fans, lights, air leakage etc.

Manage Field Temperature of the Produce

The produce after harvest is left in the field exposed to direct heat radiation, which results in 7-15°C rise in temperature of the produce compared to one kept under shade (Anonymous 1986). Since rate of respiration is governed by temperature for every 5-10°C rise in temperature of the commodity, the respiration rate roughly doubles/triples. This shaded produce with 10°C lower field temperature would result in saving in refrigeration cost by as much as 17-28 per cent (Satish *et al.*, 2004).

Reduce Heat Transmission into Store

It is assumed that of the total heat entering a cold storage room, about 70 per cent enters through roof and 25 per cent through the walls. Heat input through floor is assumed to be nil as earth remains at constant temperature throughout the year. Increasing the insulation to walls and roofs can minimize the heat input to cold storage room. Doubling the insulation reduces transmitted heat by half. Newer installations use insulation levels as high as R 60 in the ceilings and R 40 in the walls. It is advisable to build with more insulation as it is much cheaper to install insulation during construction than afterwards. Painting the exterior with light colour decreases heat flow into the storage room. A dark flat roof can be 42°C warmer than the outside air temperature. Painting south-facing wall a light colour can reduce the effective wall temperature by 11°C compared with a dark wall.

Optimise Cooling Operation

Finned coil evaporator (FCE) should be preferred over bunker or diffuser type system due to more conducting surface area of fins per unit length of pipe. FCE needs only 2-3 per cent of storage space compared to 15-20 per cent (approx. 8 feet height of store) required in the latter. Diffusers use centrifugal blowers and vertical air duct which needs very large amount of power in comparison with direct driven axial flow fan in FCE; thereby saving about 30 per cent energy. Storage installation requiring zero/sub-zero temperature should not have more than 2-3 fins per inch (FPI). For high storage temperature, a maximum of 14 FPI can be selected.

The temperature of the refrigerant fluid after it is cooled in the condenser should be as low as possible to minimize energy use. Heat exchange surface of condenser may be increased to further reduce the refrigerant condensing temperature. Evaporative condensers should be preferred over air-cooled units. The most economical water flow rates are 2.5-3 gal/min./ton for recirculated water system. The water-cooled systems are subjected to scaling of condenser tubes. Periodic use of an inhibited solution of muriatic acid (18 per cent) will keep off scaling load.

Rotary compressor is the best option for small capacity and centrifugal compressor for very large installation (50 tons or above). Reciprocating compressors operate efficiently over a large range of refrigerant flows for peak loads. For every 50 MT capacity 1 TR/hr is the best criteria for selecting the capacity of a compressor.

Energy Efficiencies of Pre-cooling Systems

Different pre-cooling methods are opted for different commodities. The energy efficiency of these systems can be compared based on energy coefficients (Table 9.12). Vacuum-cooling is the most energy-efficient followed by hydro-cooling and forced air-cooling. The reason for high efficiency of vacuum-cooling is that it removes heat only from the product being cooled (Table 9.13). The other types of coolers remove heat from fans, pumps, lights, leakages, conduction through walls and people working on the cooler. The difference in efficiency represents the potential for energy saving. A properly operated hydro-cooler can operate as efficiently as vacuum-cooler and can have nearly as low energy efficiency as a forced air-cooler.

Table 9.12: Typical Energy Efficiency in Pre-cooling Operations

Type of Cooler	*Energy Coefficients*
Vacuum-cooler	2.5–1.6
Hydro-cooling	2.3–0.7
Forced air-cooling	1.5–1.4

Source: Thompson *et al.*, 2001.

Table 9.13: Distribution of Heat Input in Pre-cooling

Item	*Percent of Total Heat Input*		
	Vacuum	*Hydro*	*Forced Air*
Product	100	54	47
Fan/pumps	0	9	37
Leakage/start-up, conduction	0	37	7
Lift trucks	0	0	8
Light/people etc.	0	0	1

Source: Thompson *et al.*, 2001.

Vacuum Cooling

Refrigeration demand starts increasing after 10 min from start to a maximum when the product temperature-drop actually begins. Reducing energy consumption and improved efficiency of a vacuum cooler can be achieved by turning off the compressor when no refrigeration is needed. Operating cooler with partial load reduces energy efficiency; a half load of lettuce requires 50 per cent more energy per carton to cool than a full load. Cooling time also affects the energy consumption *e.g.*, a load of cauliflower which cools in 2 to 3 hours takes three times more energy per kg of product cooled than iceberg lettuce that requires only 20-30 minutes.

Hydro-coolers

Protecting cooler from external heat gain can reduce energy consumption. Over 1/3 rd of heat input to a hydro-cooler is due to heat leakage, heat conducted through walls and cooling the water reservoir at start-up each day. Using plastic flap doors and minimizing distance between shower pan and top of the product can reduce cooling loss. Adding insulation or painting the cooler with light colour reduces heat conduction through walls. Using smaller water reservoir lowers the amount of start-up cooling needed. Placing the pre-cooler in a refrigerated building saves energy cost by 30 per cent.

Forced Air Cooler

Here fan energy consumption is the most significant. In addition to their own energy consumption, fans contribute over

33-50 per cent of heat that is removed by forced air-cooler. Fan energy consumption can be minimized by turning the fans off when not needed, installing evaporator coils with minimum air flow resistance, using cartons with adequate venting (5 per cent) area and aligning pallet-vents with cooler to reduce air-flow resistance. At the start of pre-cooling, fan speed required is about 1 litre/sec/kg of product stored to remove the residual field-heat of the product. But once the product has reached the desired temperature, air movement can be reduced by 60-80 per cent. Variable frequency controls are commonly used to slow fans at the end of a cooling cycle in order to reduce heat input without significantly slowing the cooling process. Shift forced air-cooling operation to night (least expensive option) if there is excess cooling capacity available at night and if products can withstand some cooling delay.

Package Icing

Many operators put more ice than is needed to cool the product. This results in wastage of ice without any changes in product quality. In a box of Broccoli, this extra represents almost half of the ice used in the cooling process. Ice systems are often poorly insulated or shaded from the sun. Table 9.14 throws a light on the comparative performance of commonly used pre-cooling methods.

Important tips

1. Harvest the crop at night or early in the day or keep harvested crop under shade during the cooling delays.
2. Each 1-hour delay between harvest and cooling results in the loss of one day of shelf-life.
3. Produce left at ambient dry conditions looses 100 times more moisture than produce shifted to a cold room.
4. For lower refrigeration cost, build cube shaped store room.
5. Prefer PUF as insulation to thermocole; R-60 for roof and R-40 for walls.
6. Operate the fan at optimum air flow rates.
7. Use moist air (RH > 95 per cent) to prevent desiccation of produce.
8. Use cartons with at least 5 per cent vent area for optimum cooling.

Table 9.14: Comparison of Typical Product Effects and Cost for Common Cooling Methods

Attribute	*Cooling Methods*				
	Room	*Forced Air*	*Hydro*	*Ice*	*Vacuum*
Cooling time(hr)	20–100	1-10	0.1-1.0	0.1–0.3	0.3–2.0
Moisture loss(per cent)	0.1–2.0	0.1-2.0	0-0.5	-	2.0-4.0
Water contact with product	No	No	Yes	Yes, unless bagged	No
Potential for decay contamination	Low	Low	High	Low	None
Capital cost	Low	Low	Low	High	Medium
Energy efficiency	Low	Low	High	Low	High
Water resistant packaging needed	No	No	Yes	Yes	No
Portable	No	Sometimes	Rarely done	Common	Common
Feasibility of online cooling	No	Rarely done	Yes	Rarely done	No

Contd...

Table 9.14–Contd...

Attribute	*Cooling Methods*				
	Room	*Forced Air*	*Hydro*	*Ice*	*Vacuum*
Recommended for	Snap beans, cabbage, carrots, brinjal, okra, peppers, cucumber, potatoes, dry onions, summer squash, reddish, garlic, tomato, turnip, apples, pears.	Snap beans, broccoli, cabbage, pepper, cucumber, brinjal, melons, okra, peas, summer squash, apples, tomatoes, strawberries, cherries, grapes, peaches, pears, plum.	Asparagus, snap beans, broccoli, Brussels sprout, cauliflower, sweet corn, cucumber, lettuce, green onions, radish carrots, green peas, spinach, turnip, apples, mangoes, apricot, cherries, pear, peach, plum	Asparagus, broccoli, Brussels sprouts, carrots, sweet corn, lettuce, melons, green onions, reddish, spinach, turnip, cantaloupe	Brussels sprouts, cauliflower, sweet corn, spinach, lettuce, cauliflower
Not recommended for		Leafy vegetables and those having low thickness of thin stem or peel	Grapes, cut flowers, melons	Peppers, okra, mango, tomatoes, Kinnow, guava and other subtropical products	Carrots, potato, tomato

Source: Thompson *et al.,* 1998, Gast and Flores 1991, Kitinoza and Gorny 1998, Hardenberg *et al.,* 1990.

9. Forced air-cooling is 4-10 times faster than room-cooling but is 2-23 times slower than hydro-cooling or vacuum-cooling.
10. Treat cooling water with recommended disinfectant (chlorine, phenols, hypochlorite) to prevent decay producing micro-organisms.
11. Hydro-cooling is slower for crated products as compared to the loose ones.
12. One kg of ice will roughly drop the temperature of product 3 times its weight by about 28°C.
13. Use laminated or wax impregnated CFB box rather than plain CFB box.
14. Vacuum-cooling is more energy efficient and economical despite high initial cost and gives more rapid and uniform cooling compared to other methods of cooling.
15. Reduce moisture loss by pre-wetting or spraying water on the produce before it is placed in vacuum-cooling-chamber.
16. According to IIR recommendations, perishable products should be brought to their final storage temperature within 24 hour of harvest.

References

Anonymous.1967. Recommended conditions for cold storage of perishable produce. 2nd ed. Int. Institute of Refrigeration.

Anonymous.1981. Vacuum cooling for fruits and vegetables. Food Processing Industry 50 (601): 24.

Anonymous.1986. Improvement of postharvest fruits and vegetable handling–A Manual. Bangkok. RAPA/FAO/AFMA.

Anonymous.1994. ASHRAE handbook: Methods of pre-cooling fruits, vegetables and cut flowers. Refrigeration Systems and Applications.

Anonymous.1998. User's manual for the research vacuum cooler. Southport, UK: Autec Ltd.

Anonymous.1982. American Society of Heating Refrigerating and Air Conditioning Engineers. ASHRAE Handbook. Applications. Chapter 59., ASHRAE, Atlanta, GA.

Arfin BB, Chau KV. 1988. Cooling of strawberries in cartons with new vent hole design. ASHRAE Transaction 92: 1415-26.

Athey D, Dennis C. 1991. Vegetable processing. Glasgow: Blackie.

Atkins F C. 1974. Guide to mushroom growing. London: Faber and Faber.

Baird CD, Gaffiney J J, Talbot MT.1988. Design criteria for efficient and cost effective forced air cooling system of fruits and vegetables. ASHRAE Transactions 92: 1434-41.

Balls RC. 1986. Horticultural engineering technology: fixed equipment and buildings. London: Macmillan.

Barger WR. 1961. Factors affecting temperature reduction and weight loss of vacuum cooled lettuce. USDA Market Res. Report 469:20.

Barger, W.R. 1962. Vacuum cooling lettuce in commercial plants. USDA, Market. Serv. AMS-469: 9.

Barger, W.R. 1963. Vacuum pre-cooling- a comparison of the cooling of different vegetables. USDA, Market. Res. Rpt. 600: 12.

Bartlett D. 1971. Vegetable review: Cooling and storage. Commercial Grower. 3950: 383-84.

Bennett, A.H. Smith, R.E., and Fortson, J.C. 1965. Hydro-cooling peaches–a practical guide for determining cooling requirements and cooling times. USDA, Inf. Bull. 293.

Boyette M.D. and E.A. Estes. 2004. Crushed and liquid ice cooling. Published by North Carolina Cooperative Extension Service, Raleigh/North Carolina Agricultural and Technical State University, Greensboro.

Boyette M.D., E.A. Estes and AR Rubin. 2004. Hydro-cooling. Published by North Carolina Cooperative Extension Service, Raleigh/North Carolina Agricultural and Technical State University, Greensboro,NC 27695-7603.

Boyette M.D., L.G. Wilson and E.A. Estes. 2004. Design of room cooling facilities: structural and energy requirements. Published by North Carolina Cooperative Extension Service, Raleigh/North Carolina Agricultural and Technical State University, Greensboro.

Boyette M D, Wilson LG, Estes EA 1994. Forced air cooing. North Carolina Cooperative Extension Service, Raleigh/North

Carolina Agricultural and Technical State University, Greensboro, AG-414-3.

Boyette M D, Wilson LG, Estes EA. 1994. Hydro-cooling. North Carolina Cooperative Extension Service, Raleigh/North Carolina Agricultural and Technical State University, Greensboro, AG-414.

Boyette M D, Wilson LG, Estes EA. 1994. Introduction to proper postharvest cooling and handling methods. North Carolina Cooperative Extension Service, Raleigh/North Carolina Agricultural and Technical State University, Greensboro, AG 414-1.

Brash D. Carrots. 1996. Postharvest handling for export. Levin Research Centre, The New Zealand Institute for Crop and Food Research Limited, Private Bag 4704, Christchurch, New Zealand.

Brosnan, T and Sun Da-Wen. 2001. Pre-cooling techniques and applications for horticultural products–a review. Int. J of Refrigeration 24: 154-170.

Buffington DE, Sastry Sk, Gustashaw JC, Buris DS. 1981. Artificial curing and storage of Florida onions. Transactions of the ASAE 24 (3): 782-88.

Cantwell M and Thangaiah A. 2001. Delays to cool affect visual quality, firmness and gloss of bell peppers and eggplants. Perishable Handling Quarterly, No. 107: 17-20.

Chen, VL. 1988. Vacuum, hydro and forced air-cooling of farm produce and their energy consumption. FFTC book series, no. 37, Taiwan pp 104-111.

Dennis C. 1984. Effect of storage and distribution conditions on the quality of vegetables. Acta Hort. 163: 85-104.

DiRisio T. 1994. Steam ejector vacuum systems keep food processors cool. Food Processing 1: 73-4.

Eaks, IL 1956. Effectivity of hydro-cooling on oranges. Citrus leaves 6 92), 2-33.

Fordham A, Biggis A.G 1985. Principles of vegetable crop production. London: Collins Professional and Technical p. 178-179.

Friedman, B.A., and W.A. Radspinner. 1956. Vacuum cooling fresh vegetables and fruits. USDA Market. Serv. AMS-107: 15.

Gast Karen L.B. and Rolando Flores. 1991. Pre-cooling produce, Fruits and Vegetables. Cooperative Extension Service. Kansas State University, Manhattan, Kansas.

Gast, L.B. Karen and Rolando A. Flores. 1992. Storage operations, fruits and vegetables. Cooperative Extension Service, Manhattan, Kansas MF-1033.

Gibbon JM. 1972. Some observations of temperature and cooling rates of vegetables in commercial cold stores. J. of Agril. Engng. Res. 17 (4): 332-7.

Gillies SL, Toivonen PMA.1995. Cooling method influences the Postharvest quality of broccoli. Hort. Science 30 (2): 313-15.

Grizzell WG, Bennett AH. 1966. Hydro-cooling stacked crates of celery and sweet corn. USDA Agril. Res. Report, pp 52.

Guillou, R. 1960. Coolers for fruits and vegetables. Calif. Agr. Exp. Sta. Bull. 773.

Guillou, R. 1970. Pre-cooling in the west. ASHRAE. Proc. of Symp. on Pre-cooling Fruits and Vegetables. SF. 4: 11-13.

Hall, E.G. 1972. Pre-cooling and container shipping of citrus fruits. *CSIRO Food Res. Quarterly* 32, 1-10.

Hardenberg, R.E; A.E. Watada and CY Wang. 1990. The commercial storage of fruits, vegetables and florist and nursery stocks. USDA, Agril. Res. Service,Handbook No. 66.

Heinze, PH. 1961. Effect of storage on potato quality. Potato Hand Book 6: 32-36.

Hughes, P.A., Thompson, A.K., Plumber, R.S. and Seymour, G.B. 1981. Storage of capsicum under controlled atmosphere, modified atmosphere and hypobaric conditions. *J. of Hor. Science* 56, 261-265.

Janick, J. 1986. Horticultural Science. 4th Ed. New York. Freeman WH, pp 550-51

Kader AA. 1987. Respiration and gas exchange of vegetable. In: Postharvest Physiology of vegetables. J. Weichmann (ed) Macel Dekker Inc. New York

Kim BS, Kim DC, Lee Se, Nahmgoong Choi MJ, Joong MC. 1995. Freshness prolongation of crisphead lettuce by vacuum cooling. Agril. Chemistry and Biotechnology 38 (3): 239-47.

Kitinoza, L and Gorny, J.R. 1998. Postharvest technology for small scale produce marketers: Economic opportunities, quality and food safety. UC Postharvest outreach programme. PHPTC, USAID/ACE project, University of California, California.

Kumar Satish, Dhatt AS and Mahajan BVC. 2004. Postharvest temperature management in horticultural crops: Effect of colour, size and shading on the pulp and surface temperature of fruits and vegetables. Paper presented and published in International Conference on Emerging technologies in agricultural processing engineering (etae 2004), held at IIT, Kharagpur on 14-17 December 2004.

Lindsey RT, Neale MA, Messer HJM. 1983. Ventilation rates for positive ventilation of vegetables in bulk bins. J. Agril. Engng. Res 28 (1): 33-34.

Lindsey, R.T. and Neale, M.A. 1977. Proceedings of the Vegetable Cooling and Storage Conference, National Agricultural Centre, Stoneleigh, UK.

Lovelidge B. 1972. Vacuum cooling backed by mechanized field packing leads to improved marketing and better use of labour. The Grower 78 (6): 239-40.

Mitchell, F.G., R. Guillou and R.A. Parsons. 1972. Commercial cooling of fruits and vegetables. Calif. Agr. Exp. Sta. Serv. Manual 43.

Neale, M.A., Lindsey, R.T. and Messer, H.J.M. 1981. An experimental cold store for vegetables. *J.l of Agril. Engng. Res.* 26, 529-540.

Nowak, J. Rudnicki R.M. 1990. Posthavest handling and storage of cut flowers florist greens and potted plants. London: Chapman and Hall.

Pantastico, B. 1975. Postharvest physiology, handling and utilization of tropical and subtropical fruits and vegetables. AVI Publishing Co., Inc., Westport, Connecticut.

Pasqualone SB. 1986. Experimental trails on the vacuum pre-cooling of various vegetables and fruit products. Rivistadi Frutticolturae-di Ortofloricoltura 48 920: 45-50.

Powell, G.H. 1905. The transportation of fruits in refrigeration. ASRE Trans. 1 pp 82-94.

Prange RK. 1994. Postharvest cooling of horticultural crops. Production technology report. Agriculture and Agri-Food Canada, Research Station, Kentville, NS, B4N1J5.

Reid MS. 1991. Effect of low temperature on ornamental plants. Acta Hort. 298: 215-222.

Reid MS, Kofranek AM. 1980. Postharvest physiology of cut flowers. Chronica Horticulturae.

Robertson J. 1978. Mechanizing vegetable production. Ipswich (Suffolk): Farming Press Limited, 1978.

Rudnucki RM, Nowak J, Goszczynska DM. 1991. Cold storage and transportation conditions for cut flower cuttings and potted plants. Acta Hort. 298: 225-30.

Ryall, A.L. and W.T. Pentzer. 1982. Handling transportation and storage of fruits and vegetables (2nd edn) Vol 2. AVI Publishing Company, Inc. Westport, Connecticut.

Sainsbury GF. 1961. Cooling apples and pears in storage rooms. USDA Market Res. Rep. p 474.

Showalter RK, Greirson W. 1972. Pre-cooling fruits and vegetables in the southeast. ASHRAE Symposium Bulletin SF-4-70, p 2.

Smith, R.E, Nelson GL, Henrickson RL. 1967. Analysis on transient heat transfer from anomalous shapes. Transaction of the ASAE 10 (2): 236-45.

Smith, W L JR. 1957. Peach hydro-cooling, shipping and fungicidal tests. USDA, AMS. 199.

Stanley R. 1989. The influence of temperature and packaging material on the postharvest quality of iceberg lettuce. Acta Hort 244: 171-7.

Sullivan GH, Davenport LR, Julian JW. 1996. Progress in new crops: Pre cooling: key factor for assuring quality in new fresh market vegetable crops. Arlington (VA): ASHS Press, p. 521-524.

Thompson, JF. 1992. Storage systems. In: Kader, A A(ed). Postharvest technology of horticultural crops. UC,Davis, Division of Agriculture and Natural Resources Publication.3311.

Thompson J, Cantwell M, Arpaia ML, Kader A, Crisosto C and Smilanick J. 2001. Effect of cooling delays on fruit and vegetable quality. Perishable Handling Quarterly Issue No. 105: 2-5.

Thompson J. 2001. Energy conservation in cold storage and cooling operations. Perishable Handling Quarterly, No. 105: 7-9.

Thompson JF, Chen YL. 1988. Comparative energy use of vacuum, hydro and forced air coolers for fruits and vegetables. ASHRAE Transactions 92: 1427-33.

Thompson JF, Kesmire RF, Mitchell FG. 1998. Room, vacuum, transport and ice cooling, commercial cooling of fruits, vegetables and flowers. Publication 21567. University of California, Division of Agriculture and Natural Resources.

Thompson, A.K. 1996. Postharvest technology of fruits and vegetables. Oxford Blackwell Science.

Thompson, J.F. 1998. Pre-cooling and storage facilities. Department of Biological and Agricultural Engineering, University of Calfornia, Davis, USA.

Turk R, Celik E. 1994. The effects of vacuum cooling on the quality criteria of some vegetables. Acta Hort 368: 825-9.

Van Blaricom, L O. 1959. The effect of decay of adding various reagents to the water for hydro-cooling peaches. S. C.Agr. Exp. Sta. Circ.124.

Wang CY. 1993. Approaches to reduce chilling injury of fruits and vegetables. Hort. Rev. 15: 63-95.

Winkler AJ, Cook JA, Kliewer WM, Lider LA. 1974. General iticulture. Berkley: University of California Press.

Zemansky MW, Abbott MM, Van Ness HC. 1975. Basic engineering thermodynamics. 2nd ed McGraw-Hill, New York.

Chapter 10

Process Optimization of Cereal-Banana Based Ready to Eat Extruded Snack Food

K. Karthika, K. Thangavel and R. Viswanathan

Department of Food and Agriculture Process Engineering, Agriculture Engineering College & Research Institute, TNAU, Coimbatore – 3 (T.N.)

Now a days many ready to eat products have flooded the market. Inspite of diminishing area and production, the cereals continue to have an important place in Indian Agriculture. Rice is the stable food of more than 60 per cent of world's population. It is a high-energy or high-calorific food. Rice flour is an important ingredient for many ready to eat break fast cereals and snacks. Several studies (Altomane and Ghossi, 1986; Hseih *et al.*, 1993; Janes and Guy, 1995) have been conducted on the extrusion of rice flour based products. Maize is an important cereal crop in world's agricultural economy. It is a miracle crop and also called the "Queen of cereals". Ilo and Berghofer (1999) conducted studies on the utilization of yellow maize grits under extrusion cooking in a conical, counter-rotating twin-screw extruder. In view of the large intake, cereals are also an important source of several other nutrients like protein, calcium, iron and B-complex vitamins.

Fruits constitute an important part of human diet for supplying the essential vitamins, minerals and fibre. Banana is considered as the second fruit crop of tropical and subtropical regions of the world. It is a delicate and highly perishable fruit having a very special place in diet which is low in fats, cholesterol and high in calorific value. India being a tropical country, postharvest losses are much higher due to high temperature and also due to lack of technology for handling and processing (Wasker, 1996). Perishable commodities like banana suffer enormous post harvest damage and more often these losses go unaccounted particularly in the Indian subcontinent. The Indian banana growers are loosing around 20 to 60 per cent of their production on account of the existing handling practices (Salunke and Desai, 1998).

Extrusion cooking provides an opportunity to create new products and improve the existing one. Harper (1981) defined that extrusion cooking is a versatile, efficient way of converting raw ingredients into intermediate or finished foods and is used commercially to produce a great volume and variety of food products. It is used to convert indigestible cereals into products that are digestible and palatable. Chen *et al.* (1991) defined that extrusion cooking is a process where food materials are subjected to higher temperature and pressure in combination with shearing stress. This process of high-temperature short-time extrusion brings gelatinization of starch, denaturation of proteins and inactivation of enzymes and microbes.

The extruded products often lead to imbalanced nutrition among the consumers. The functional effects of some of the cereals and fruits are quit useful in changing many ready to eat products to provide balanced nutrient. There is still lack of knowledge on extrusion cooking characteristics and extrudate properties using cereal banana blend.

Hence the present study was conducted to prepare the extruded snack food with rice and maize flour by incorporating banana pulp at different levels to optimize the process parameters of extrusion cooking of banana blended with rice and maize flour, to evaluate the various quality characteristics of extrudate produced from banana blended with rice and maize flour and to study the organoleptic properties of banana blended extrudate.

Raw Material

Banana, rice and maize flour were procured from the local

market. The study was carried out with two varieties of banana viz, poovan and robusta.

Preparation of the Sample

The cereal flour banana blend for extrusion cooking was made from rice flour, maize flour and banana pulp. The rice flour and maize flour were allowed to pass through ISS sieve number 100. Banana was made into pulp and mixed with rice/maize flour using kitchen type mixer. The TSS of the banana pulp was maintained at 23 to 24° brix throughout the experiment. After blending with banana, blend flour was allowed to pass through ISS sieve number 100 three times. Moisture content of the cereal flour banana pulp blend was determined by drying at 130±1°C in a thermostatically controlled ventilated hot air oven for two hours. (AOAC, 1997). The samples were conditioned to have a compact volume by hand pressing and kept in a container for one hour prior to extrusion.

Extrusion of Flour Banana Blend

A laboratory model twin-screw extruder was used in the extrusion experiments. Through preliminary extrusion trials, the process variables were fixed as; barrel temperature- 80, 90 and 100°C, screw speed- 350, 375 and 400 rpm for rice- banana blend flour and barrel temperature of 120°C, 130°C and 140°C, screw speed of 375, 400 and 425 rpm for maize- banana blend flour. In addition to this banana pulp was added to rice and maize flour at 0,10 and 20 per cent level.

Extrudate Quality

Extrudate Quality for Cereal- banana Blend Extrudates

Expansion Ratio

Expansion ratio is taken as the index of increase in volume of a product due to extrusion cooking. It is expressed as the ratio between the extrudate diameter of the product and diameter of the die. A vernier caliper (0.02 mm least count) was used to measure the diameter of extruded samples collected from each run.

The effect of process variables on the expansion ratio of rice banana extrudates is shown in Table 10.1. It was observed that the maximum expansion ratio of 2.66 was achieved at the barrel temperature of 100°C, screw speed of 400 rpm for the rice flour

Table 10.1: Effect of Barrel Temperature, Screw Speed and Addition of Banana Pulp on Expansion Ratio for Rice-banana Extrudates

Treatments	*Temperature, °C*	*Screw Speed, rpm*	*Expansion Ratio*	
			Rice-poovan	*Rice-robusta*
Rice flour + 0% pulp			2.39	2.39
Rice flour +10% pulp		350	2.26	2.25
Rice flour +20% pulp			2.13	2.14
Rice flour + 0% pulp	80		2.41	2.41
Rice flour +10% pulp		375	2.38	2.31
Rice flour +20% pulp			2.19	2.16
Rice flour + 0% pulp			2.44	2.44
Rice flour +10% pulp		400	2.39	2.34
Rice flour +20% pulp			2.23	2.23
Rice flour + 0% pulp			2.43	2.43
Rice flour +10% pulp		350	2.41	2.39
Rice flour +20% pulp			2.27	2.26
Rice flour + 0% pulp	90		2.64	2.64
Rice flour +10% pulp		375	2.64	2.55
Rice flour +20% pulp			2.31	2.31
Rice flour + 0% pulp			2.65	2.65
Rice flour +10% pulp		400	2.64	2.59
Rice flour +20% pulp			2.32	2.30
Rice flour + 0% pulp			2.65	2.65
Rice flour +10% pulp		350	2.64	2.64
Rice flour +20% pulp			2.33	2.33
Rice flour+ 0% pulp	100		2.64	2.64
Rice flour +10% pulp		375	2.64	2.65
Rice flour +20% pulp			2.39	2.41
Rice flour + 0% pulp			2.66	2.66
Rice flour +10% pulp		400	2.63	2.65
Rice flour +20% pulp			2.42	2.41

		SED	CD (0.05)	CD (0.01)
Rice-poovan	T × S × B	0.01705	0.03417	0.04554
Rice-robusta	T × S × B	0.01738	0.03484	0.04644

without addition of banana pulp. This was closely followed by extrudates obtained from rice flour without banana pulp at the temperature of 100 °C with screw speed of 350 rpm and 90°C with a screw speed of 400 rpm. The minimum value of expansion ratio, 2.13 was obtained for rice- poovan extrudate and 2.14 for rice–robusta extrudates at the temperature of 80°C with screw speed of 350 rpm blended with 20 per cent banana pulp.

The observations on expansion ratio in maize–banana extrudates are furnished in Table 10.2. From the table, it is evident that the highest expansion ratio obtained for maize-banana extrudates were 2.69 at the temperature of 130 °C, screw speed of 375 rpm and 10 per cent addition of banana pulp and the same expansion ratio was obtained at the 140°C, screw speed of 425 rpm and 10 per cent banana pulp addition. The lowest expansion ratio 2.24 was obtained at the temperature of 120 °C, screw speed of 375 rpm and 20 per cent banana in maize poovan extrudates and 2.13 obtained at the temperature of 140 °C, screw speed of 425 rpm and 20 per cent addition of banana pulp in maize robusta extrudates.

At higher temperature, due to increase in the gelatinization the expansion ratio of the extrudates increases with more air space and hence the extrudate becomes less denser. The finding was in conformation with the results obtained by Bardie and Mellowes (1991), who reported an increase in expansion of cassava flour extrudate from 2.27 to 2.34 when the extrusion temperature was raised from 100°C to 120°C.

Mean comparison of Least Significant Difference (LSD) indicates that expansion ratio at different screw speed was significantly different. This might be due to the more shearing action at higher screw speeds of the extruder. It leads to more expansion of the product with more pore space. Battacharya and Hanna (1985) observed a similar trend for corn at high screw speeds. Increase in the level of banana pulp level leads to lower the expansion ratio of the extrudates due to poor gelatinization and hence the extrudates becomes more denser.

True Density

True volume of extrudates was determined using Pycnometer (Micrometeristics, USA: model 1305). The true density was calculated as the ratio of the mass of the sample to its true volume.

Table 10.2: Effect of Barrel Temperature, Screw Speed and Addition of Banana Pulp on Expansion Ratio for Maize-banana Extrudates

Treatments	*Temperature, °C*	*Screw Speed, rpm*	*Expansion Ratio*	
			Maize-poovan	*Maize-robusta*
Maize flour + 0% pulp		375	2.60	2.60
Maize flour +10% pulp			2.58	2.59
Maize flour +20% pulp			2.24	2.27
Maize flour + 0% pulp			2.62	2.62
Maize flour +10% pulp	120	400	2.59	2.60
Maize flour +20% pulp			2.25	2.29
Maize flour + 0% pulp			2.64	2.64
Maize flour +10% pulp		425	2.61	2.63
Maize flour +20% pulp			2.26	2.30
Maize flour + 0% pulp		375	2.68	2.68
Maize flour +10% pulp			2.69	2.69
Maize flour +20% pulp			2.37	2.30
Maize flour + 0% pulp		400	2.67	2.67
Maize flour +10% pulp	130		2.65	2.66
Maize flour +20% pulp			2.27	2.31
Maize flour + 0% pulp		425	2.64	2.64
Maize flour +10% pulp			2.64	2.70
Maize flour +20% pulp			2.26	2.32
Maize flour 0% pulp			2.63	2.63
Maize flour +10% pulp		375	2.55	2.63
Maize flour +20% pulp			2.43	2.26
Maize flour + 0% pulp	140		2.64	2.64
Maize flour +10% pulp		400	2.57	2.62
Maize flour +20% pulp			2.51	2.13
Maize flour + 0% pulp			2.65	2.65
Maize flour +10% pulp		425	2.69	2.69
Maize flour +20% pulp			2.28	2.13

		SED	CD (0.05)	CD (0.01)
Maize-poovan	T × S × B	0.02371	0.04753	0.06335
Maize-robusta	T × S × B	0.00642	0.01286	0.01714

The effect of processing parameters on true density is shown in Table 10.3. From the table, it was inferred that the maximum value of true density of 1446.33 kg/m^3 at the barrel temperature of 100°C, screw speed of 400 rpm and zero per cent addition of banana pulp were obtained for both varieties of banana blended extrudates. The minimum true density of 1139.33 kg/m^3 for rice-poovan extrudate and 1135 kg/m^3 for rice robusta extrudates was obtained at the temperature of 80°C with screw speed of 350 rpm and 20 per cent banana pulp level.

The true density obtained during extrusion cooking of maize-banana extrudates is given in the Table 10.4. It was observed that the true density was maximum at the temperature of 140°C with screw speed of 400 rpm and 10 per cent banana level for both varieties of maize-banana extrudates. True density varied from 1165.33 kg/m^3 to 1398.00 kg/m^3 for maize–poovan extrudates and for maize-robusta extrudates, true density varied from 1171.67 kg/m^3 to 1402.00 kg/m^3. Mean comparison by LSD reveals that temperature, screw speed and addition of banana pulp had a significant effect on the extrudate true density.

Cutting Strength

The force necessary to cut the sample into pieces was determined by using an Instron Universal Testing Machine with a Warner Bratzler blade. The samples were analysed at a speed of 0.05 mm/s, load cell 100 N and sample length of 6 mm.

The observations on cutting strength during extrusion cooking at different processing parameters in rice-banana extrudates are presented in the Figures 10.1 and 10.2. It was inferred that the minimum cutting strength of 0.38 N/mm^2 was achieved at the temperature of 100°C, screw speed of 400 rpm, and zero per cent banana pulp for both varieties of banana blended extrudates. The maximum value for cutting strength 0.98 N/mm^2 for rice–poovan extrudates and 0.97 N/mm^2 for rice-robusta extrudates was observed at the temperature of 80°C, screw speed of 350 rpm and 20 per cent banana pulp.

The observation on cutting strength during extrusion cooking at different processing conditions in maize- banana extrudates are presented in Table 10.5. From the table, it was inferred that the maximum value of 0.99 N/mm^2 for maize-poovan extrudates and 0.96 N/mm^2 for maize- robusta extrudates was obtained at the

Table 10.3: Effect of Barrel Temperature, Screw Speed and Addition of Banana Pulp on True Density for Rice-banana Extrudates

Treatments	Temperature, °C	Screw Speed, rpm	True Density, Kg/m³	
			Rice-poovan	Rice-robusta
Rice flour + 0% pulp			1306.00	1306.00
Rice flour +10% pulp		350	1234.00	1236.67
Rice flour +20% pulp			1139.33	1135.00
Rice flour + 0% pulp	80		1347.33	1347.33
Rice flour +10% pulp		375	1262.67	1286.00
Rice flour +20% pulp			1193.00	1197.33
Rice flour + 0% pulp			1394.67	1394.67
Rice flour +10% pulp		400	1306.33	1307.00
Rice flour +20% pulp			1205.67	1215.33
Rice flour + 0% pulp	90		1347.67	1347.67
Rice flour +10% pulp		350	1266.00	1259.33
Rice flour +20% pulp			1166.33	1181.00
Rice flour + 0% pulp			1364.00	1364.00
Rice flour +10% pulp		375	1284.00	1315.00
Rice flour +20% pulp			1216.33	1224.67
Rice flour + 0% pulp			1416.33	1416.33
Rice flour +10% pulp		400	1345.00	1308.00
Rice flour +20% pulp			1264.00	1277.33
Rice flour + 0% pulp			1362.33	1362.33
Rice flour +10% pulp		350	1295.00	1258.33
Rice flour +20% pulp			1202.00	1270.00
Rice flour + 0% pulp	100		1382.67	1382.67
Rice flour +10% pulp		375	1307.67	1318.33
Rice flour +20% pulp			1242.00	1282.67
Rice flour + 0% pulp			1446.33	1446.33
Rice flour +10% pulp		400	1385.33	1310.00
Rice flour +20% pulp			1309.00	1325.33

		SED	CD (0.05)	CD (0.01)
Rice-poovan	T × S × B	3.93779	7.89480	10.52128
Rice-robusta	T × S × B	20.89874	41.89952	55.83881

Table 10.4: Effect of Barrel Temperature, Screw Speed and Addition of Banana Pulp on True Density for Maize-banana Extrudates

Treatments	*Temperature, °C*	*Screw Speed, rpm*	*True Density, Kg/m³*	
			Maize-poovan	*Maize-robusta*
Maize flour + 0% pulp		375	1262.00	1262.00
Maize flour +10% pulp			1213.67	1218.33
Maize flour +20% pulp			1165.33	1171.67
Maize flour + 0% pulp	120	400	1292.33	1292.33
Maize flour +10% pulp			1272.33	1279.00
Maize flour +20% pulp			1196.00	1212.67
Maize flour + 0% pulp		425	1325.67	1325.67
Maize flour +10% pulp			1314.67	1317.33
Maize flour +20% pulp			1216.33	1221.33
Maize flour + 0% pulp		300	1306.67	1306.67
Maize flour +10% pulp			1283.33	1283.00
Maize flour +20% pulp			1205.33	1218.67
Maize flour + 0% pulp		400	1333.33	1333.33
Maize flour +10% pulp	130		1314.67	1316.33
Maize flour +20% pulp			1222.00	1234.67
Maize flour + 0% pulp		425	1356.00	1365.00
Maize flour +10% pulp			1333.33	1339.00
Maize flour +20% pulp			1234.33	1243.33
Maize flour + 0% pulp	140	375	1365.00	1365.00
Maize flour +10% pulp			1347.00	1356.67
Maize flour +20% pulp			1234.67	1240.00
Maize flour + 0% pulp		400	1394.00	1394.00
Maize flour +10% pulp			1398.00	1402.00
Maize flour +20% pulp			1261.33	1267.33
Maize flour + 0% pulp		425	1393.00	1393.00
Maize flour +10% pulp			1395.67	1396.67
Maize flour +20% pulp			1268.00	1287.67

		SED	CD (0.05)	CD (0.01)
Maize-poovan	T × S × B	2.91724	5.84872	7.79450
Maize-robusta	T × S × B	2.84439	5.70266	7.59985

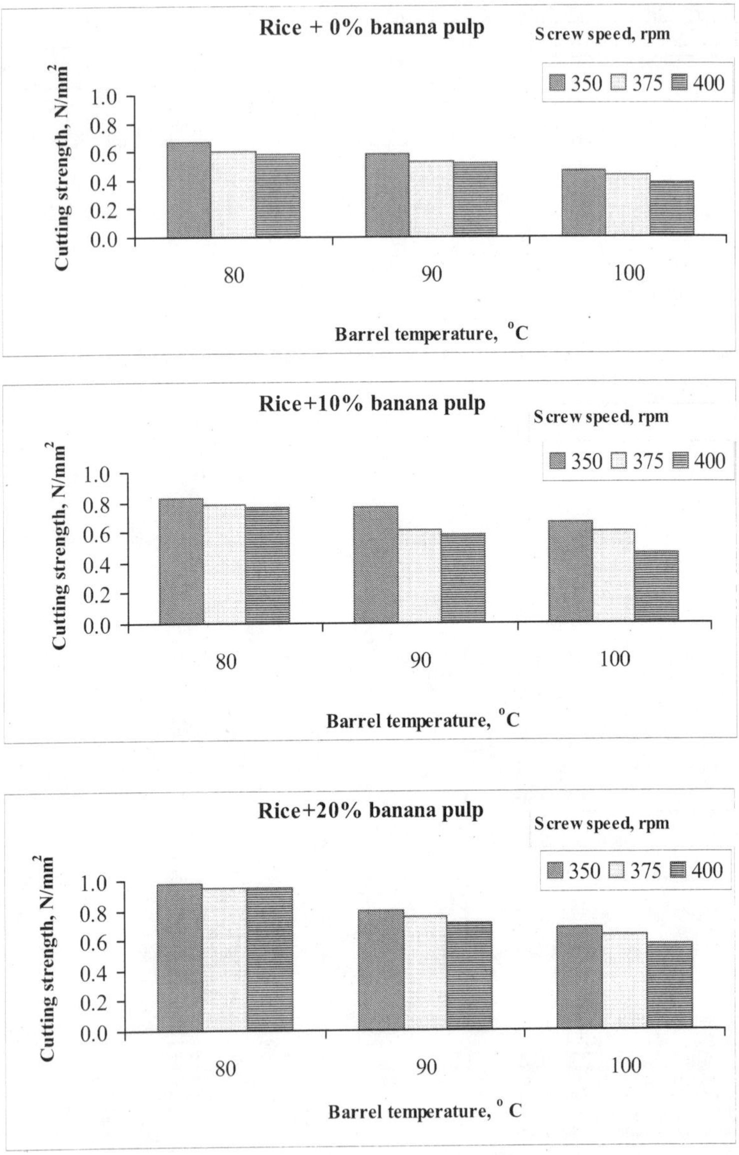

Figure 10.1: Effect of Barrel Temperature, Screw Speed and Addition of Banana Pulp on Cutting Strength of Rice-poovan Extrudates

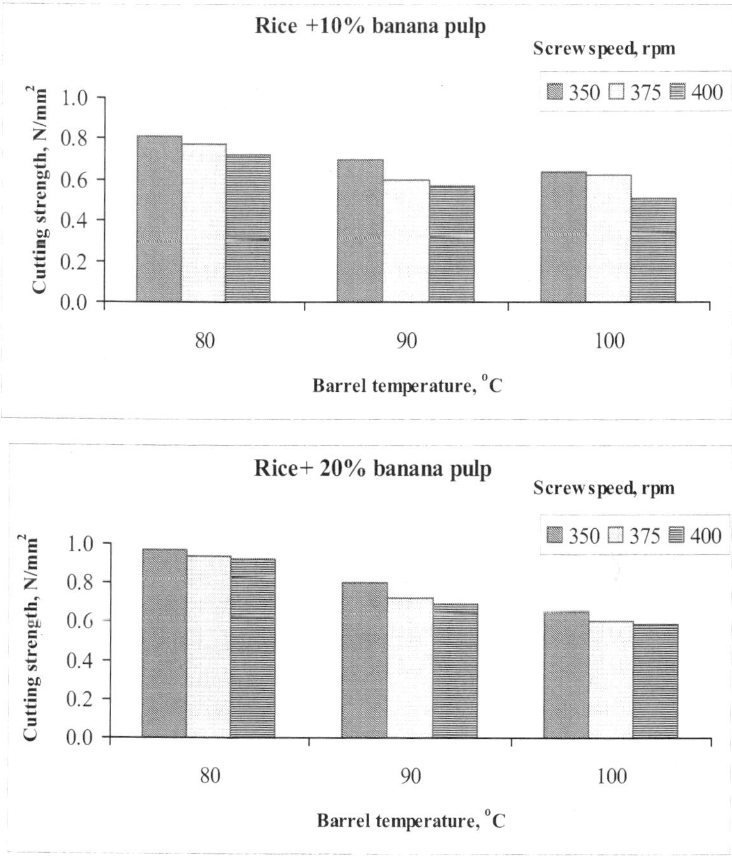

Figure 10.2: Effect of Barrel Temperature, Screw Speed and Addition of Banana Pulp on Cutting Strength of Rice-robusta Extrudates

		SED	CD (0.05)	CD (0.01)
Rice-poovan	T × S × B	0.0497	0.00996	0.01328
Rice-robusta	T × S × B	0.00559	0.01121	0.01494

temperature of 120°C, screw speed of 375 rpm and 20 per cent banana pulp level. The minimum value of 0.41 N/mm² for maize-poovan extrudates and 0.42 N/mm² for maize-robusta extrudates was obtained at the temperature of 140°C, screw speed of 425 rpm and 10 per cent banana pulp level.

Table 10.5: Effect of Barrel Temperature, Screw Speed and Addition of Banana Pulp on Cutting Strength for Maize-banana Extrudates

Treatments	Temperature, °C	Screw Speed, rpm	Cutting Strength, N/mm²	
			Maize-poovan	Maize-robusta
Maize flour + 0% pulp			0.71	0.71
Maize flour +10% pulp		375	0.83	0.85
Maize flour +20% pulp			0.99	0.96
Maize flour + 0% pulp	120		0.65	0.78
Maize flour +10% pulp		400	0.80	0.83
Maize flour +20% pulp			0.96	0.93
Maize flour + 0% pulp			0.61	0.66
Maize flour +10% pulp		425	0.77	0.80
Maize flour +20% pulp			0.94	0.89
Maize flour + 0% pulp			0.60	0.62
Maize flour +10% pulp		375	0.71	0.75
Maize flour +20% pulp			0.81	0.87
Maize flour + 0% pulp	130		0.58	0.60
Maize flour +10% pulp		400	0.68	0.71
Maize flour +20% pulp			0.78	0.83
Maize flour + 0% pulp			0.59	0.59
Maize flour +10% pulp		425	0.67	0.68
Maize flour +20% pulp			0.70	0.78
Maize flour + 0% pulp			0.53	0.54
Maize flour +10% pulp		375	0.57	0.64
Maize flour +20% pulp			0.67	0.74
Maize flour + 0% pulp	140		0.47	0.47
Maize flour +10% pulp		400	0.42	0.51
Maize flour +20% pulp			0.59	0.67
Maize flour + 0% pulp			0.42	0.42
Maize flour +10% pulp		425	0.41	0.42
Maize flour +20% pulp			0.51	0.63

		SED	CD (0.05)	CD (0.01)
Maize-poovan	T × S × B	0.00737	0.01478	0.01969
Maize-robusta	T × S × B	0.02753	0.05520	0.07356

At higher temperature, due to increase in the gelatinization, the extruded product became more porous which required low force to cut the extrudate into two pieces when compared to low level of the gelatinzation at low temperature. The increase in screw speed would increase the shear rate, which tends to decrease cutting strength of the product. Lo *et al.* (1998) reported a reduction in breaking strength from 7.79 to 4.13 N while the screw speed increased from 370 to 447 rpm. The reduction in breaking strength was attributed to the screw speed in decreasing the cell wall thickness.

Water Absorption Index (WAI)

These were determined by according to the procedure laid down by Jin *et al.* (1995).

The effect of process parameters on water absorption index on rice–banana extrudates is shown in Table. 10.6. The maximum water absorption index of 7.57 g/g was observed in rice–poovan extrudates and 7.60 g/g in rice–robusta extrudates. The minimum value was 5.03 g/g for both varieties of banana-blended extrudates.

The minimum water absorption index maize–banana extrudates of 5.03 g/g was obtained at the temperature of 140°C, screw speed of 425 rpm and zero per cent banana pulp addition for both varieties of banana blended extrudates. The maximum value of 7.43 g/g for maize-poovan and 7.07g/g for maize-robusta extrudates was obtained at the temperature of 120°C, screw speed of 375 rpm and 20 per cent banana pulp addition (Figures 10.3 and 10.4).

Temperature had an inverse effect on water absorption index. The same trend was obtained by Kaur *et al.* (2000) for rice grits extruded with mustard oil. Increased screw speed resulted in decrease in the water absorption index This might be due to mechanical disruption of starch granules at high screw speed. Lo *et al.* (1998) reported that increase in the screw speed resulted in reduction of water absorption index from 3.05 to 1.79 g/g.

Water Solubility Index (WSI)

Table 10.7 shows the water solubility index. The index ranged from 29.33 to 55.67 per cent for both varieties of banana-blended extrudates.

Table 10.8 indicated that maximum water solubility index of 51.33 per cent was obtained at the temperature of 140°C, screw speed

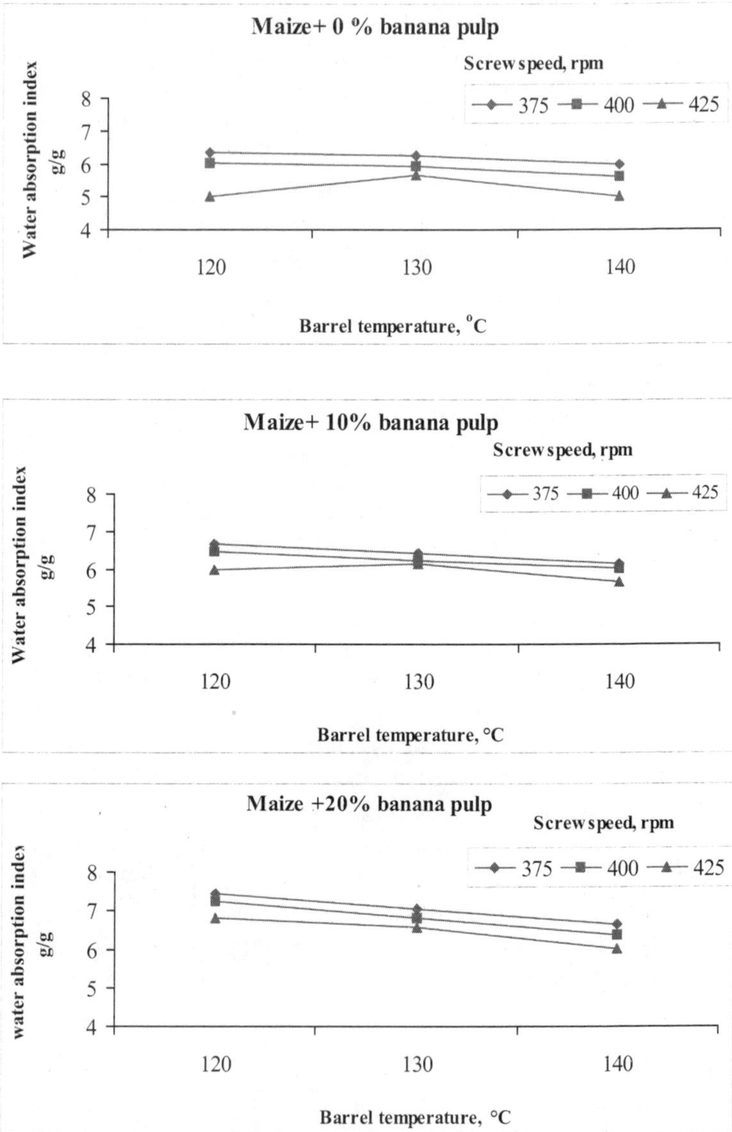

Figure 10.3: Effect of Barrel Temperature, Screw Speed and Addition of Banana Pulp on Cutting Strength of Rice-robusta Extrudates

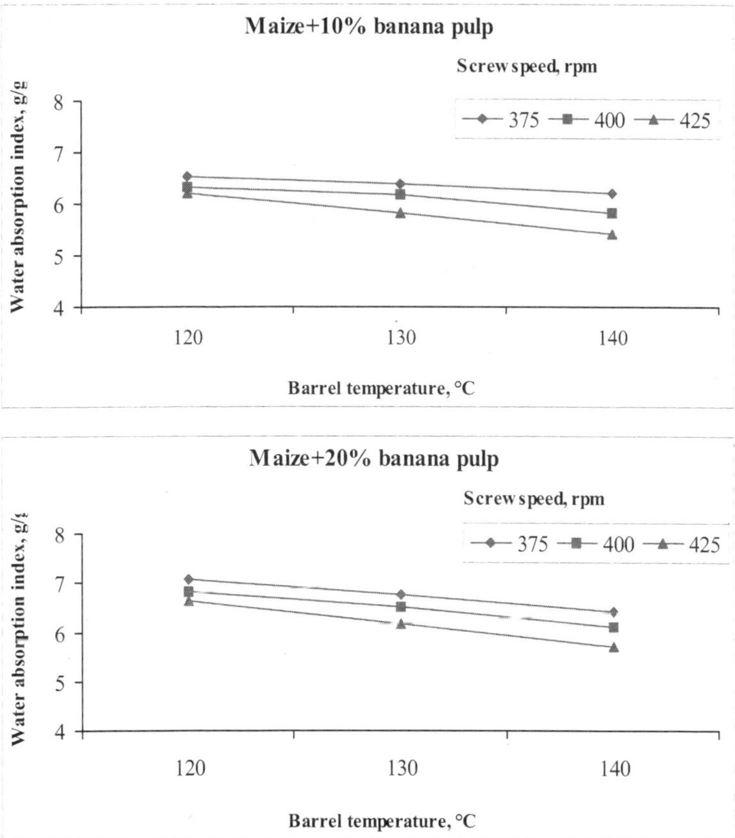

Figure 10.4: Effect of Barrel Temperature, Screw Speed and Addition of Banana Pulp on Water Absorption Index of Maize-robusta Extrudates

		SED	CD (0.05)	CD (0.01)
Maize-poovan	T × S × B	0.04444	0.08911	0.11875
Maize-robusta	T × S × B	0.05290	0.10606	0.14134

of 425 rpm and zero per cent banana pulp level for both varieties of banana blended extruded products. The minimum value of 34.30 per cent for maize-poovan and 34.97 per cent for maize-robusta extrudates was obtained at the temperature of 120°C, screw speed of 375 rpm and 20 per cent banana pulp level.

Table 10.6: Effect of Barrel Temperature, Screw Speed and Addition of Banana Pulp on Water Absorption Index for Rice-banana Extrudates

Treatments	*Temperature, °C*	*Screw Speed, rpm*	*Water Adsorption Index, g/g*	
			Rice-poovan	*Rice-robusta*
Rice flour + 0% pulp			6.90	6.90
Rice flour +10% pulp		350	7.10	7.20
Rice flour +20% pulp			7.57	7.60
Rice flour + 0% pulp			6.80	6.80
Rice flour +10% pulp	80	375	7.03	7.20
Rice flour +20% pulp			7.37	7.50
Rice flour + 0% pulp			6.53	6.53
Rice flour +10% pulp		400	6.97	7.13
Rice flour +20% pulp			7.20	7.43
Rice flour + 0% pulp			6.63	6.63
Rice flour +10% pulp		350	6.83	6.80
Rice flour +20% pulp			7.33	7.30
Rice flour + 0% pulp			6.40	6.40
Rice flour +10% pulp	90	375	6.83	6.73
Rice flour +20% pulp			7.20	7.20
Rice flour + 0% pulp			5.63	5.63
Rice flour +10% pulp		400	6.47	6.57
Rice flour +20% pulp			6.63	6.67
Rice flour + 0% pulp			6.23	6.23
Rice flour +10% pulp		350	6.43	6.37
Rice flour +20% pulp			7.03	7.03
Rice flour + 0% pulp			6.00	6.00
Rice flour +10% pulp	100	375	6.33	6.23
Rice flour +20% pulp			6.87	6.77
Rice flour + 0% pulp			5.03	5.03
Rice flour +10% pulp		400	6.20	6.13
Rice flour +20% pulp			6.43	6.47

		SED	CD (0.05)	CD (0.01)
Rice-poovan	T × S × B	0.05518	0.11064	0.14744
Rice-robusta	T × S × B	0.07751	0.15540	0.20710

Table 10.7: Effect of Barrel Temperature, Screw Speed and Addition of Banana Pulp on Water Solubility Index for Rice-banana Extrudates

Treatments	Temperature, °C	Screw Speed, rpm	Water Solubility Index, %	
			Rice-poovan	Rice-robusta
Rice flour + 0% pulp			37.00	37.00
Rice flour +10% pulp		350	34.64	35.67
Rice flour +20% pulp			29.33	29.33
Rice flour + 0% pulp			42.33	42.33
Rice flour +10% pulp	80	375	37.67	38.00
Rice flour +20% pulp			32.67	33.00
Rice flour + 0% pulp			47.67	47.67
Rice flour +10% pulp		400	45.33	40.67
Rice flour +20% pulp			36.33	35.33
Rice flour + 0% pulp			44.33	44.33
Rice flour +10% pulp		350	40.33	40.33
Rice flour +20% pulp			35.00	34.67
Rice flour + 0% pulp			47.67	47.67
Rice flour +10% pulp	90	375	42.67	43.67
Rice flour +20% pulp			37.33	38.00
Rice flour + 0% pulp			50.00	50.00
Rice flour +10% pulp		400	46.33	47.67
Rice flour +20% pulp			40.33	41.00
Rice flour + 0% pulp			50.67	50.67
Rice flour +10% pulp		350	45.33	45.00
Rice flour +20% pulp			40.33	40.67
Rice flour + 0% pulp			52.67	52.67
Rice flour +10% pulp	100	375	47.67	46.33
Rice flour +20% pulp			42.33	42.33
Rice flour + 0% pulp			55.67	55.67
Rice flour +10% pulp		400	49.33	51.00
Rice flour +20% pulp			46.00	44.67

		SED	CD (0.05)	CD (0.01)
Rice-poovan	T × S × B	0.66046	1.32415	1.78468
Rice-robusta	T × S × B	0.75903	1.52177	2.02804

Table 10.8: Effect of Barrel Temperature, Screw Speed and Addition of Banana Pulp on Water Solubility Index for Maize-banana Extrudates

Treatments	Temperature, °C	Screw Speed, rpm	Water Solubility Index, %	
			Maize-poovan	Maize-robusta
Maize flour + 0% pulp	120	375	40.43	40.43
Maize flour +10% pulp			38.27	39.27
Maize flour +20% pulp			34.30	34.97
Maize flour + 0% pulp		400	42.47	42.33
Maize flour +10% pulp			40.47	41.63
Maize flour +20% pulp			37.37	36.23
Maize flour + 0% pulp		425	44.37	44.37
Maize flour +10% pulp			42.33	43.57
Maize flour +20% pulp			39.53	38.40
Maize flour + 0% pulp	130	375	44.53	44.53
Maize flour +10% pulp			41.60	42.60
Maize flour +20% pulp			38.47	39.43
Maize flour + 0% pulp		400	47.10	46.43
Maize flour +10% pulp			43.40	42.67
Maize flour +20% pulp			45.57	40.73
Maize flour + 0% pulp		425	48.57	48.57
Maize flour +10% pulp			45.50	44.63
Maize flour +20% pulp			42.33	43.40
Maize flour + 0% pulp	140	375	47.33	47.33
Maize flour +10% pulp			42.33	42.73
Maize flour +20% pulp			40.33	41.53
Maize flour + 0% pulp		400	49.43	49.43
Maize flour +10% pulp			44.60	45.47
Maize flour +20% pulp			42.40	42.63
Maize flour + 0% pulp		425	51.33	51.53
Maize flour +10% pulp			47.33	47.50
Maize flour +20% pulp			45.07	43.80

		SED	CD (0.05)	CD (0.01)
Maize-poovan	T × S × B	0.34700	0.69570	0.92715
Maize-robusta	T × S × B	0.24360	0.48839	0.65087

The temperature has positive effect on water solubility index of rice-extrudes. The same trend was supported by Lo *et al.* (1998). It was reported that water solubility index increased from 33.1 to 45.2 per cent with an increase in barrel temperature from 144 to 172°C for corn meal.

As the screw speed increased from 350 to 400 rpm, a steady increase in water solubility index values was observed. This might be due to the mechanical disruption of starch granules at high screw speeds which made the product more porous for water absorption.

Lo *et al.* (1998) also reported an increase in water solubility index from 38.1 to 43.7 per cent corresponding to the increase in screw speed of 370 to 477 rpm for corn meal.

Selection of Optimum Combination of Variables

The desired quality parameters of the extruded products depend on higher expansion ratio, true density and water solubility index and lower values of cutting strength and water absorption index. Out of these parameters, expansion ratio was identified as the major quality parameter. The best combinations were $T_{100} S_{400} A_0$ for rice-banana extrudates and $T_{130} S_{375} A_{10}$ and $T_{140} S_{425} A_{10}$ for maize banana extrudates.

Sensory Evaluation

The extrudates obtained by blending the banana pulp at 0, 10 and 20 per cent levels with rice and maize flour at a combination which gave the maximum expansion ratio were tested for organoleptic attributes. All the samples were served to the judges under ambient conditions, during the test periods and the samples were displayed in separate booths. Nine point hedonic scale was used as sensory evaluation score card to bring out the inherent characteristics of a particular extruded product. The results were analyzed statistically.

Mean sensory scores of extruded samples from rice flour and maize flour blended with banana pulp are presented in Table 10.9

It can be construed that all the samples scored more than the mean scores of 6.45. Extrudates prepared without addition of banana pulp scored the maximum when blended with rice. However, with addition of 10 per cent banana pulp scored the maximum when blended with maize flour.

Table 10.9: Mean Scores of Sensory Evaluation of Extruded Products

Sensory Parameters	*Rice-banana Pulp,%*			*Maize-banana Pulp,%*		
	0	*10*	*20*	*0*	*10*	*20*
Colour	7.70	7.10	6.50	7.89	7.80	6.56
Appearance	7.20	7.15	6.45	8.00	8.10	6.21
Flavour	7.20	7.00	6.45	7.68	7.69	7.15
Crispiness	6.90	6.80	6.40	7.58	7.63	6.53
Taste	7.20	7.15	6.45	7.56	7.58	7.35
Overall acceptability	7.55	7.45	6.65	7.90	7.91	6.68
Mean scores	**7.29**	**7.11**	**6.48**	**7.77**	**7.79**	**6.75**

From the study it can be concluded that the optimum level of process parameters, *viz;* barrel temperature and screw speed were 100°C and 400 rpm for rice–banana blend, and barrel temperature of 140°C with screw speed of 425 rpm for maize banana extrusion studies. The desired quality parameters *viz.* expansion ratio, true density, cutting strength, water absorption and water solubility index for rice- banana extrudates at optimum levels of process parameters were 2.66, 1446.33 kg/m^3, 0.38 N/mm^2, 5.03 g/g and 55.67 per cent. Similarly, at optimum combination of variables the quality parameters of maize–poovan extrudates were 2.69, 1395.67 kg/m^3, 0.41N/mm^2, 5.67 g/g and 47.33 per cent. For maize–robusta extrudates, the quality parameters of 2.69, 1396.67 kg/m^3, 0.42N/mm^2, 5.40 g/g and 47.50 per cent were obtained. Sensory evaluation studies revealed that rice flour without banana pulp and 10 per cent banana pulp blended with maize resulted in highly acceptable extruded product. From this study, it is seen that banana can be extruded along with cereals. This can increase the utilization of banana in addition to its use for table purposes.

References

*Altomare, RE., and P. Ghossi.1986. An analysis of residence time distribution patterns in a twin screw cooking extruder. Biotechnol Prog., 2: 157-163.

AOAC.1997. Official Methods of Analysis. (Eds). Association of Official Analytical chemists. Washington, DC.

Bardie, N and W.A.Mellowes.1991. Effect on extrusion variables on cassava extrudates. J. Food Sci., 56 (5): 1334-1337.

Bhattacharya, M. and M.A. Hanna.1985. Extrusion processing of wet corn glutted meal. J. Food Sci., 50:1508-1509.

Chen, J., F.L. Serafin, N.P. Rajesh and H.Daun.1991. Effects of extrusion conditions on sensory properties of corn meal extrudates. J.Food Sci., 56 (1): 84-89.

Harper, J.M.1981. Extrusion of foods. VolI. CRC press, Boca Raton, Florida.

Hsieh, F., K.M.Grenus and H.E. Hulf.1993. Twin screw extrusion of rice flour with salt and sugar, Cereal Chem., 70: 493-498.

Ilo, S., and E. Berghofer.1999. Kinetics of colour changes during extrusion cooking of maize grits. Journal of Food Engg., 39: 73–80

Janes, D.A. and R.C.E.Guy.1995. Metastable sates in a food extrusion cooker-II. The effects of die resistance and minor ingredients with rice flour. J. Food Eng.26: 161-175.

Jin, Z., F.H. Siech and H.E. Haff.1995. Effect of soy fibre, salt, sugar and screw speed on physical properties and microstructure of corn meal extrudates. J. Cereal Sci., 22: 185-194.

Kaur, L., N.Singh, K.Kaur and B.Singh. 2000. Effect of mustard oil and process variables on extrusion behavior of rice grits. J. Food Sci.Technol., 37(6): 656-660.

Lo, T.E.R.G. Moreira, M.E. Castell-Perez.1998. Modeling product quality during twin-screw food extrusion. Transactions of the ASAE, 41 (6): 1729-1738.

Salunke, D.K. and D.B. Desai.1998. Post harvest disease of horticulure perishables. International book distributing company. Lucknow. India.

Waskar, D.P and S.K. Roy.1996. Methods adopted for extending the storage life of banana fruit. A review in Agric. Res.,17: 132-138.

* Original not consulted.

Chapter 11

Packages for Export of Horticultural Produce

S.C. Mandhar and G. Senthil Kumaran

Indian Institute of Horticultural Research, Hessaraghatta, Bangalore – 560 089

India is one of the major producers of fruits and vegetables. However, its share in export market is negligible. One of the reasons for the meagre export is the unavailability of standard packaging techniques. For increasing the export potential of these perishable commodities, internationally acceptable packaging norms need to be followed. Corrugated Fibre Board (CFB) containers are accepted internationally in the trade for export of most of the horticultural produce. The size of the CFB container should be such that these can be palletized and mechanically handled. The strength of the container should be appropriate to take care of handling, stacking and transportation loads. Proper cushioning and isolations are to be provided to reduce damage to the product due to stacking and vibrations. Proper evaluation of the container is a prerequisite for its use. The containers are to be properly labelled for contents, grade and quality of produce, the name of producer/exporter and the country of origin.

Design and Test Requirements of Packages

Size of Packages

Corrugated fibre board containers of fully-telescopic and regular-slotted types are used for export of horticultural produce.

Packages designed for conveying pallets should be of the size of 1200 × 1000 mm or 1200 × 800 mm to permit efficient mechanical handling. Interlockable sizes of packages (600 × 400, 600 × 200, 520 × 240, 500 × 333, 433 × 333, 400 × 300, 380 × 240 and 300x200 mm) are preferred over modular stacking sizes. Consumer size packages are preferred in the trade where as the master container should be designed to adequately contain consumer size packages. Size of the package should be just sufficient to hold the product tightly. Excess empty space results in vibrational damage and excessive transpiration losses. Size of package is generally optimized based on cost and strength. Packages with single layer of fruits are preferred for mango and pine apple.

Requirements for Corrugated Fibre Board

All stacking and handling loads need to be taken up by the CFB. The requisite number of plies, grammage and type of flutes are designed based on the above need. In India, kraft paper has to meet the requirements of the Bureau of Indian Standards IS:1397-1967 (BIS, 1967). The CFB has to meet bursting strength requirements as per IS:2771 (Part-I)-1977(BIS, 1977), puncture resistance as per IS:4006 (Part-II)-1985 (BIS, 1985) and cobb value as per IS:4006 (Part-I)-1985 (BIS, 1985). The container should have preferably ISI marking IS:2771 (Part-I).

Strength Requirement of Container

Most fruits and vegetables are pre-cooled, packaged and stored under conditions of low temperature and high humidity. Accordingly, the boxes are to be designed for 0°C and 100 per cent relative humidity (RH). For transporting leafy vegetables, the packages are to be preferably wax coated from inside. Height of stacking in most aircrafts and trucks is around 2.5 m. Packages are designed to withstand 3 to 5 times the normal stacking height to take the load due to shock and vibrations. It may be advantageous to make sure prior to purchase that the flutes of the side walls are vertical. Small-sized boxes are preferred because of enhanced compression strength per unit area for similar specifications of CFB. Fully-telescopic containers (FTC) are preferred over regular-slotted containers (RSC) for getting more strength because of the provision of double walls on all the sides of the containers.

The packages are classified into two groups according to their ability to withstand stacking and storage up to a maximum height of (i) 2.5 m: group A, and (ii) 8 m: group B, as per the recommendations of the Organization for Economic Co-operation and Development (OECD).

Ventillation

Ventilation holes, when required should be such that the area covered by the holes should not exceed 5 per cent of the total surface area of the package. Excessive number of holes reduces the strength of the package and increase transpiration losses. Ventilation holes near the corners should be avoided.

Isolation and Cushioning

Individual big sized fruits are isolated and cushioned in packages using:

1. Corrugated fibre board cell pack
2. Pulp tray with or without styro-foam sheet cushion
3. Styro-foam cell tray pack.

The consumer size packages of small fruits are usually isolated using corrugated fibre-board cell pack. Onion, cabbage, carrot and some leafy vegetables don't require any isolation and cushioning.

Presentation and Labelling of Containers

The outer ply of CFB boxes should preferably be of white and thick duplex-board. Descriptive marking as per Organisation for Economic Co-operation and Development (OECD) should be printed on the container. The basic contents of the label are:

1. Name and address of packer and/or dispatcher
2. Nature of produce
3. Variety (commercial name)
4. Country of origin
5. Trade name (where appropriate)
6. Size (where appropriate)
7. Weight, number of units (where appropriate) and

8. Grade/class of produce.

Testing and Evaluation of Filled Containers

Packages are tested under simulated transport conditions as per one of the following two standards:

1. American Society for Testing and Materials, Standard Practice for Performance Testing of Shipping Containers and Systems as per ASTM D4169-86 (ASTM, 1986).
2. Organisation for Economic Co-operation and Development Recommendation on the Standardization of Packaging for International Trade.

As per ASTM D4169-86, cycle of handling and transport is simulated through drop tests for manual/mechanical handling, compression tests for warehouse stacking and vehicle stacking and vibration tests for vibrations during transport. Each handling, stacking and transport operation during the distribution cycle is simulated separately as it occurred in transport and distribution cycle. Individual containers are given six drops at the surfaces, edges and corners from a height of 17.8 to 22.9 cm as per ASTM D775-80 (ASTM, 1980) to simulate manual handling. To simulate mechanical handling of palletized containers, two edge drops are given from a height of 7.6 to 15.2 cm as per ASTM D1083-85 (ASTM, 1985). The containers are conditioned at 0°C and 100 per cent RH for 24 hours. Filled containers are tested for 3 times the stacking load on the lower most container as per ASTM D642-47 (ASTM, 1976) to simulate dynamic load. Filled containers are dwelled at 0.5 g for 4 resonant frequencies for 5 mm as per ASTM D999-75 (ASTM, 1981) to simulate transport vibrations. The damage to the product should be within acceptable limits.

At the Indian Institute of Horticultural Research, 19 packages were tested in total as per OECD recommendations. Five packages were tested for compression, 9 for vibration and 5 for free fall tests. The test samples were conditioned to 20±2°C and 90±2 per cent for 48 h as per ISO 2233. Empty packages were tested for 1.5 times the stacking load on lower most container for different grades of containers as per ISO 2872 and 2874. The vertical deformation should not be more than 5 per cent or 10mm. Vibration tests were conducted on 3 loaded packages placed on top of each other and additional dead load was placed on these to simulate stack load as per ISO 2247. The stacked packages were vibrated for 10 minutes each for 9mm horizontal and 9mm vertical displacement at 4 Hz. The test was replicated thrice. Free fall tests were conducted on 5 filled containers as per ISO 2248. Two drops were given to each container on the bottom-end edge and the bottom-side edge. The packages were considered to be satisfactory if there was no permanent deformation. The test samples in the present study failed as per ASTM standards. Since the OECD recommends a higher drop height than ASTM standards for drop tests, the test samples would also not be able to pass OECD recommendations.

References

A.S.T.M. 1976. Standard Method for Compression Test for Shipping Containers, ASTM D642-47. American Society of Testing and Materials, Philadelphia, USA.

A.S.T.M. 1980. Standard Test Method for Drop Test for Loaded Boxes, ASTM D775-80. American Society of Testing and Materials, Philadelphia, USA

A.S.T.M. 1981. Standard Test Methods for Vibration Testing of Shipping Containers, ASTM D999-75. American Society of Testing and Materials, Philadelphia. USA.

A.S.T.M. 1985. Standard Method for Testing Mechanical Handling of Unitized Loads and Large Shipping Containers, ASTM D1083-85, American Society of Testing and Materials, Philadelphia, USA.

A.S.T.M. 1986. Standard Practice for Performance Testing of Shipping Containers and Systems, ASTM D4 169-86. American Society of Testing and Materials, Philadelphia, USA.

B.I.S. 1967. Indian Standard Specification for kraft paper, IS: 1397-1967. Bureau of Indian Standards, New Delhi.

B.I.S. 1977. Indian Standard Specification for corrugated fibre board boxes, IS: 2771 (Partl)-1977. Bureau of Indian Standards, New Delhi.

B.I.S. 1985. Indian Standard Method of test for paper and paper based packaging I and II)-1985. Bureau of Indian Standards, New Delhi.

Chapter 12

Machinery for Raw-Mango Processing and Export of Mango

S.C. Mandhar, G. Senthil Kumaran, A. Carolin Rathinakumari and C. Nehru

Section of Agricultural Engineering, Indian Institute of Horticultural Research, Hessaraghatta, Bangalore – 560 089

E-mail: mandhar@iihr.res.in

Mango is one of the most important horticultural crop of India, accounting for about 42 per cent of the area under fruit cultivation. An area around 1.1 million hectares is under mango cultivation and 9.45 million tonnes of mango is produced in the country where as roughly 1 per cent of the produce is processed for different products. (Pruthi, 1992). Raw-mango pickle, chutney and powder are the delicacies of Indian subcontinent and about 9.5 per cent of the processed mangoes are used for pickle production, 3.6 per cent for mango slices in brine and 3.2 per cent for mango *chutney*. Unpeeled raw-mangoes are used for preparing pickle and peeled raw-mangoes are required for *chutney*, *amchoor* and pickle. Export of pickles and *chutneys* (including mango preparation) was Rs.564.3 million during 1996-97 (Bansal, 1998).

The flow diagram for a raw-mango processing industry is shown in Figure 12.1. The raw-mango peeling, grating and cutting is done manually and is labour-intensive. The pickle industry with

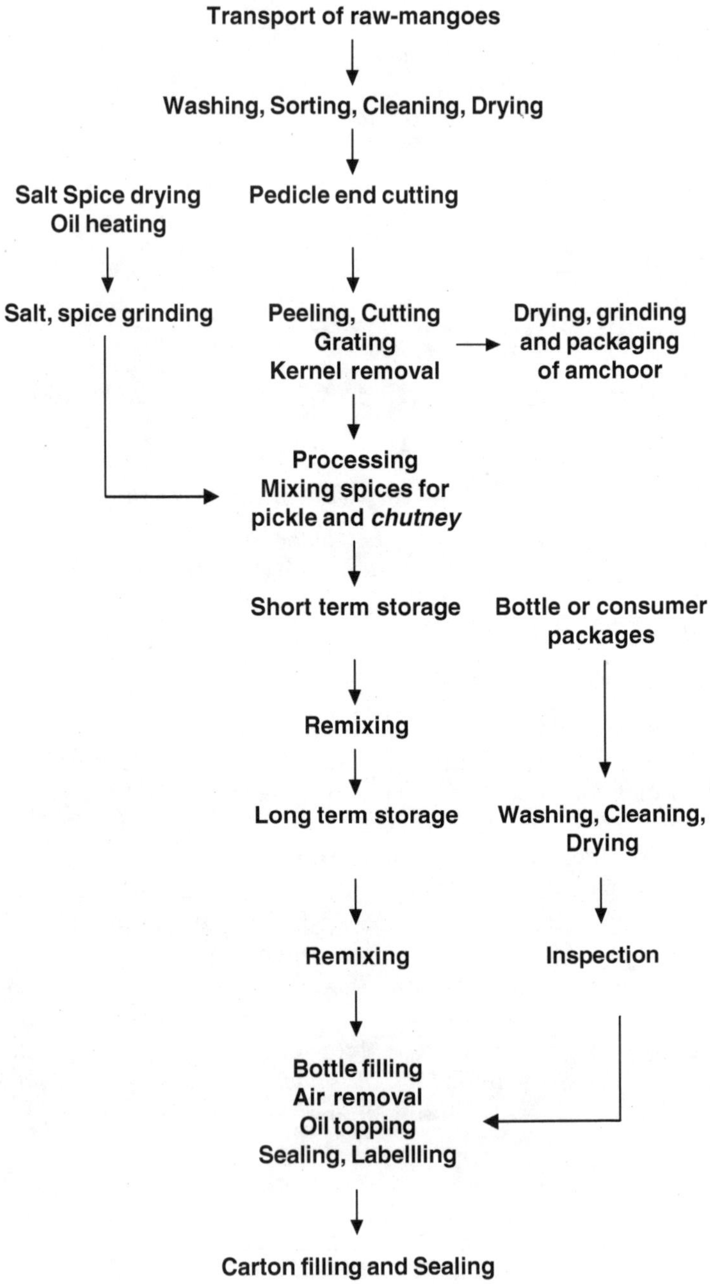

Figure 12.1: Flow Chart for Raw-mango Processing

20 tonnes per day capacity employs upto 2000 workers. As most of the operations are manual, proper hygienic conditions cannot be maintained in the industry. The operations like grinding of salt, spices and oil mixing are usually mechanized. As the raw-mango processing season lasts for about 2 months, the necessity of mechanization of the labour intensive operations has been felt. Raw-mango peeler, grater, slicer and cube cutter have been developed. Since grading is one of the prerequisites for mechanization, hence the mango grader has also been developed.

Machinery for Raw-mango Processing

Raw-mango Grader

The mangoes are made to roll under gravity (Figure 12.2) from the feeding end of the grader in the expanding open channel

Figure 12.2: Raw-mango Grader

(Mandhar, 1995). Three grades separated are based on the opening-size and the fourth consists of over-flown mangoes. The openings for grade I, II and III are 50-57; 57-64 and 64-70 mm respectively. The slope of the grader can also be adjusted for easy rolling of different varieties. The testing was conducted for *Alphonso, Totapuri, Neelam* and *Chinna Rasam* varieties. The grader can have a capacity upto 2 tonnes/hr and it does not need any electrical power and costs around Rs. 20,000.

Raw-mango Peeler

The peeler as shown in Figure 12.3 (Mandhar, 1995) consists of the frame, power transmission mechanism, concave wheel, guide drum, casing and the necessary components made of stainless steel. The inlet for un-peeled mangoes into the concave wheel has sharp projections and revolves at 250 rpm, with 2 hp motor. The guide drum revolves at 2 rpm and maintains continuous feeding and output of the mangoes. The rotation of mangoes in the elliptical chamber formed between the concave wheel (with sharp projections), guide and the casing peels the mangoes. The peeler can be modified to a grater by changing the shape of the concave wheel. The machine has to be fitted with 2 hp electrical motor for 200 kg/hr capacity. The cost of the peeler is Rs 50,000 approximately.

Figure 12.3: Raw-mango Peeler

Raw-mango Slicer

The mango-slicer (Figure 12.4) slices the mangoes longitudinally. The mangoes are cut into 4, 6 or 8 slices based on the size. The width of the slice is 20 mm and its thickness depends upon pulp-thickness whereas the length is as per the length of the mango. The mangoes are cut by the blades made out of AISI 420 grade hardened stainless steel (cutlery grade). The feeding of mangoes to the cutting blades is under gravity and a particular orientation is required to facilitate cutting of mangoes into desired sized slices. The cut slices fall on the nylon-mesh and the cut pieces of kernel are separated from the pulp under impact. The cut slices and kernel pieces are collected on one side in containers of 50 litre capacity.

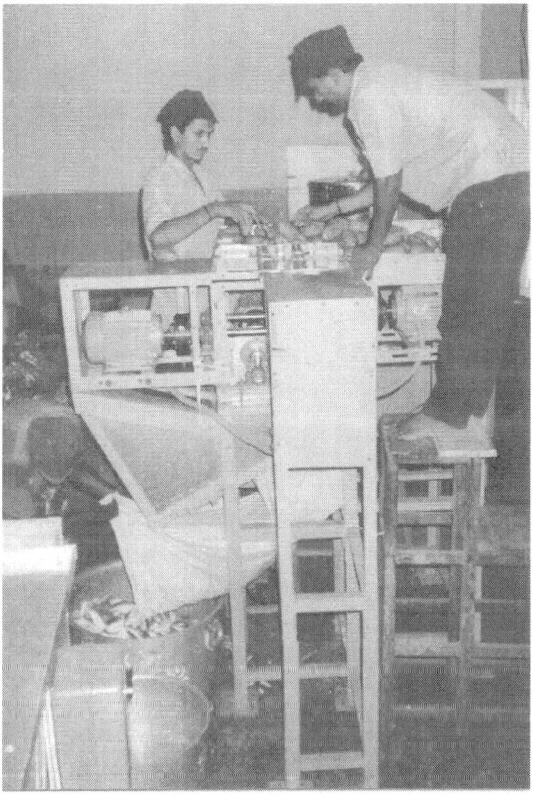

Figure 12.4: Raw-mango Slicer

The power requirement of the slicer is 6 hp for a capacity of 0.8 tonne/hr. The cost of the slicer is around Rs 60,000.

The slices and cut kernels are separated manually at present. The slices are inspected for damage and size. The wrapper/embryo membrane is removed from the endocarp manually. Further work on mechanical separation of cut kernel pieces and the slices by sieving is in progress.

Raw-mango Cube-cutter

The slices are passed through cube-cutter (Figure 12.5) and cut into cubes of approximately 2.5 cm × 2.0 cm × 1.6/1.8 cm. The blades are fabricated out of AISI 420 grade hardened stainless steel (cutlery grade). The other parts of the cube-cutter are fabricated out of AISI 304 grade stainless steel. The outlet consists of nylon-net and the

Figure 12.5: Raw-mango Cube-cutter

cut-cubes fall in 50 kg containers. All the precautions are taken for safety of the workers. The machine is powered by 5 hp motor for a capacity of 0.5 tonne/hr and it costs around Rs 1,35,000.

Machinery for Export of Mango

Hot Water Treatment Plant

A commercial hot water treatment plant was designed and developed. To control fruit fly and anthracnose, the mangoes were treated with hot water. A stainless steel tank of size 2.4 m × 1.1 m × 0.5 m was fabricated (Figure 12.6). Three heating coils (each of 5 kW) were provided to heat the water. A thermostat with digital temperature indicator was mounted to control the temperature.

Electronic Weight Grader for Mango

An electronic weight grader having capacity of 350 kg per hour for a single channel was designed, fabricated and tested. The weight-grader consisted of a fruit singulation unit, load cell, fruit grading unit, and control unit. (Figure 12.7). The conveyor belt type singulation unit fitted at a slope of 45° to the horizontal and running

Figure 12.6: Hot Water Treatment Plant for Mango

Figure 12.7: Electronic Weight Grader for Mango

at a speed of 0.2 metre per second carried one fruit at a time and fed into the load cell.

A fruit-pushing mechanism fitted over the load cell pushed the fruit into the grading unit at a speed of 24 fruits per minute. The grading unit was fitted with two gates to divert the over-weight and under-weight fruits into the corresponding outlet. When a fruit was fed into the load cell from the singulation unit, the control unit received the signal and processed these into three weight ranges and the corresponding relay switch in the control unit was activated.

The relay switches in turn operated the corresponding gate in the grading unit through solenoid switches. When the weight of the fruit on the load cell was in the required weight range say, 300 to 450 grams, no gate would operate and the fruits were collected in the good-fruit outlet. When the weight of the fruit was not in the required weight range the over-weight or under-weight fruit gates would operate accordingly and these fruits got collected in the respective outlet. This grader is very useful for small scale mango export units and fruit processing industries. It can also be used for other fruits like apple, orange, pomegranate, guava and sapota etc. with little or no modification.

Performance of the Machinery

Machinery for Raw-mango Processing

The raw-mango grader was tested for *Alphonso, Totapuri, Neelam* and *Chinna rasam* varieties at 8°, 12° and 16° slope. Grading efficiency is given in Table 12.1. It was found that the grader performed better at 12° slope with 84 per cent grading efficiency on an average. This grading efficiency is suitable for further mechanising the operations of peeling and slicing. The capacity of the grader can be upto 2 tonnes per hour depending upon the number of channels and the variety.

The continuous motorized raw-mango peeler of 200 kg per hour capacity (with one concave of 900 mm diameter) was tested for *Neelam, Alphonso, Langra, Raspuri* and *Totapuri* varieties. The performance was compared with manual peeling operation and is given in Table 12.2. The peeling efficiency was 95 to 98 per cent. The man-hours required for peeling one tonne mango per shift manually were 55 to 133 whereas it required around 5 to 9 man-hours for peeler. The capacity of the peeler can vary from 0.2 to 1.0 tonne per hour depending upon the size and number of the concave wheels provided in the machine. Upto 3 concave wheels can be provided in the machine to handle all sizes of the graded mangoes.

Table 12.1: Performance of the Mango Grader

Mango Variety	*Mean % of Correct Sizes Graded for All Openings*		
	Angle		
	8°	*12°*	*16°*
Totapuri	82.7	00.1	78.2
Alphonso	83.3	84.9	79.1
Neelam	85.1	86.0	81.4
Chinna Rasam	80.5	82.8	76.8

Table 12.2: Performance of Mango-peeler

Parameter	*Manual Peeling*	*Mango Peeler*
Capacity, kg/hr/person	7.5–18	106–200*
Labour required(man-hrs/tonne of mango)	55–103	5–9*
Weight of Un-peeled mango (gram)	144–280	142–244
Weight of peeled mango (gram)	130–210	118–175
Recoverable pulp (% of total weight of mango)	61–72	56–64
Peel removed as % of total weight of mango, (%)	10–21	18–30
Peeling Efficiency (%)	100	95–98
Cost of peeling (Per tonne of mango) (Labour cost @ Rs 40/day)	Rs 280-667	Rs 26-47
Cost of pulp production (per tonne of mango) (Mango @ Rs 4/Kg)	Rs 6209–7012	Rs 6320-7228

* The capacity of the peeler can be upto 500 kg/hr/person, the labour requirement can be upto 2 man-hours per tonne for a peeler of capacity 1 tonne/hour.

The capacity of the slicer was found to be around 0.7 tonnes per hour based upon slicing 100 tonnes of mangoes. The performance of the machine is given in Table 12.3 and Table 12.4. The slicing-efficiency based on the required size of the slices was around 90 per cent. It required two persons for feeding the mangoes with an output of 5.5 tonnes per shift of 8 hours. The labour requirement for slicing mango on the slicer was only 18.7 per cent of what is required for manual-slicing in the form of rings. It was noticed that around 90 per cent of cut pieces of kernel were separated in the slicer. The cut pieces of kernels were separated from the slices manually and the wrapper/embryo membrane was removed from the endocarp by a sharp-edged tool.

Table 12.3: Performance of Mango Slicer for *Chinna Rasam* Variety

Labourers required for feeding (nos.)	2
Capacity (kg/hr)	653–687
Percentage of slices with endocarp (per cent)	82–90
Usable slices with mechanical damage (per cent)	7–14

The raw-mango cube-cutter was tested using the raw-mango slices sliced by raw-mango slicer (Tables 12.4–12.6). The capacity of the machine was around 0.4 tonne per hour. The slices were fed manually by 4 labourers. In one shift, the output of the machine was around 3 tonnes. The labour requirement for cube cutting machine is 13.3 per cent as compared to manual cube cutting. The cost of cube cutting by the machine was Rs.194/tonne in comparison to Rs.352/tonne for manual cube cutting. The total labour required for slicing and cube cutting of one tonne mangoes on the machines was 13 man-hours (14.8 per cent) in comparison to 88 man-hour for manually cutting the cubes. The cubes and gratings were produced more hygienically on these machines.

Table 12.4: Performance of the Cube-cutter for *Chinna Rasam* Variety

Labourers required for feeding (nos.)	4
Capacity (kg/hr)	331–397
Percentage of cubes with endocarp and in the required weight range (per cent)	62–72
Usable cubes for mango paste (per cent)	7–14

Table 12.5: Comparison of Labour Requirement for Machines and Manual Operation for One Tonne of Mango

Item	*Manual*		*Slicer and Cube-cutter*	
	Ring Cutting	*Cube Cutting*	*Longitudinal Slicing*	*Cube Cutting*
Man-hours required	16	72	3	10
Total (Slicing + Cubing) man-hour	88		13	

The raw-mango grader was tested in the Agricultural Engineering Unit, University of Agricultural Sciences, Bangalore. Raw-mango Peeler tested at the Section of Agricultural Engineering, Indian Institute of Horticultural Research, Bangalore. Raw-mango Slicer and Raw-mango Cube-cutter were tested at Ushodhaya Entreprises Limited (Priya Food Division), Vijayawada.

Table 12.6: Comparison of Cost of Operation for Machines and Manual Operation for One Tonne of Mango

Item	*Manual*		*Slicer and Cube-cutter*	
	Ring Cutting	*Cube Cutting*	*Longitudinal Slicing*	*Cube Cutting*
Labour (Rs)	1.6	7.2	0.3	1.0
Machine cost (Rs)				
1. Fixed Cost				
a. Depreciation [6% operational month (Rs)]	—	—	18	24
b. Interest				
[(9% operational month (Rs)]	—	—	17	23
2. Variable Cost				
a. Electricity (Rs)	—	—	21	24
b. Blades and maintenance (Rs)	—	—	5	10
Total cost/tonne for the individual operation (Rs)	64	288	73	121
Total cost for all operations of cutting mango pieces for pickle (Rs)	3.52		194	

Mechanisation of Remaining Operations

The main problem faced is regarding bottle-filling as the pickle should have required ratio of cut pieces and *masala* (spices) mix. To maintain this proportion and to remove the entrapped air in pickle, the bottles are filled manually and ramming is done. These processes can be mechanised and the entrapped air can be removed by centrifuging the bottles. The work on these aspects is under progress. The work on mechanisation of stem-end cutting after grading is also in progress.

References

Bansal, R.K. 1998. Fruits and vegetables processing industry in India, Indian Food Packer, 52(5): 33-35.

Mandhar, S. C and Senthil Kumaran, G. 1995. Development of a continuous motorized peeler for raw-mangoes, J. Food Sci. Technol 32(1): 65-67.

Mandhar, S. C and Senthil Kumaran, G. 1995. Development of mango grader. Proceeding of National Seminar on Postharvest Technology of Fruits. 7-9, August, Bangalore, pp: 40-42.

Pruthi, J.S. 1992. Simple innovation in mango processing technologies. Ind. Food Packer 46(5): 45-55.

Chapter 13

Priorities for Postharvest Management of Agriculture and Allied Sectors in North-eastern Region

K.K. Satapathy

Division of Agricultural Engineering,
ICAR Research Complex for NEH Region, Umroi Road,
Umiam – 793 103 (Meghalaya)

North-eastern India comprising of the states of Assam, Arunachal Pradesh, Meghalaya, Manipur, Mizoram, Nagaland and Tripura, situated between 20° 05' to 29° 30' N latitude and 89° 46' to 97° 03' E longitude represents a distinct agro-climatic region. It covers an area of 255 thousand square kilometers, which is about 8 per cent of the country's total area. Except patches of plain lands, the region is locked-up by the surrounding hills with elevation ranging from 0 to 5000 m above the mean sea level. The terrain is extremely difficult and is geographically isolated with wide variations in slopes and altitudes, diverse agro-climatic conditions (ranging from sub tropical to alpine), land tenure systems and cultivation practices. The transport and communication system is poorly developed. The majority of the areas in the region still remain inaccessible and the infrastructural facilities like supply of input, marketing, institutional credit and extension services are inadequate. Majority of the population is dependent on agriculture, horticulture and allied land-based activities. However, productivity of the various crops is much lower as compared to other areas of the country. The region also experiences hot and humid weather, which is highly hazardous from the postharvest technology point of view. By virtue of geographical position, north-eastern region receives very high rainfall and the weather is highly susceptible for the development of micro-organisms. Due to high rainfall, the sunshine hours are not effective to dry-up agricultural commodities and the total postharvest losses of food grains are much higher than that in the other parts of the country. The facilities for postharvest operations such as drying, threshing, storage and processing have practically not been developed. Also, lack of proper market also acts as disincentive. Perishable commodities like fruits and vegetables have farther problems in storage, packaging, transportation and marketing.

Traditional Postharvest Technology

Agricultural processing methods are rather more crude in north-east hill region as compared to those prevalent in other parts of the country. There are however, traditional postharvest technologies that need to be understood from scientific and technological angle so as to take full advantage from their strengths and rationalize/ standardize them wherever feasible. The postharvest operation such as drying, storage, threshing and transportation of produce are highly labour intensive. Among the various postharvest operations, storage of food products and transportation are more important in the region. Since agriculture is a seasonal activity, the produce has to be kept in reserve for round the year consumption.

Drying and Storage of Agricultural Produce

Drying of grains in the north-eastern region is a major problem for summer rice which is harvested during the monsoonal period (June-July) owing to the difficulty in drying for threshing and storage. Heaping moist harvested-crop leads to moulding and spoiling of grains. In many cases the grains also start sprouting while being attached to the straw. The postharvest management in such cases poses such a problem that it becomes difficult for the small farmers to handle the crop. The indigenous method of drying followed by the farmers who cultivate the crop is to hang the harvested bunches of paddy on a horizontal bamboo (fixed on other bamboo-poles) mostly in the field itself. However, drying under such conditions is dependent on cessation of rainfall and occasional sunshine hours. If area under summer rice has to be increased, some scientific methods of drying must be offered to the farmers so that they may handle the crop properly after harvest.

Moisture, humidity, insects and rodents offer major constraints in storage. The damage to stored grains and other products due to these factors lead to reduction of food quality and hence the market value. There has been a very limited study on the losses of grains during storage in the north-eastern region. However the study undertaken by Assam Productivity Council and Assam Agricultural University, Jorhat indicated that primarily loss during storage of paddy was caused due to infestation of stored grain pests and rodents. There were also some losses due to discoloration of grains owing to the storage of grains with high moisture content. Average infestation of paddy by stored-grain pests was found to be around 4.38 per cent after 3 to 9 months of storage. The rodent-infestation has been estimated in to be around 3 to 5 per cent. Infestation of stored grain pest in maize during storage has been estimated to be around 4 per cent. However, there is practically no loss of maize due to rodent infestation because of the manner in which it is stored. The indigenous practices followed by the farmers of the NE Region for storing of rice are almost similar in most of the areas although there are some minor variations. The most common method of storing rice for consumption or sale is to have a separate storehouse with raised wooden or bamboo platform (floor). In the hills, usually unhusked paddy after threshing is stored while in the plains of Assam, rice is stored after shelling. The Assam Agricultural University has designed and recommended three types of storage structures: wooden bins, *Juria Duli* and *Lakhmi Bharal* through modification of the indigenous and existing models. The durability of the bins has been found to be 15 to 20 years. Evaluation of the bin indicated that there was no deterioration of grain quality even when paddy was stored up to a moisture content of around 15 to 16 per cent. Most of the farmers store maize as cobs in un-husked condition. In the indigenous practice of storing maize, un-husked cobs are fastened on to a frame made of bamboo poles and kept at a certain height to protect the cobs from rain. Maize cobs can be safely stored for a period of about 9 months in this way. Storage of maize is comparatively easier when preserved with the husk since the outer husk provides protection from insects and moisture etc.

Traditional Storage Structures for Food Grains

The region has still sustained conventional ways of production, preservation and processing of food. The popular structures for paddy and maize storage in Manipur are *Ningei, Dei, Kot* and *Apu achouba*. *Chuyak yums* are used for temporary storage of maize (Table 13.1). Naga tribes build their storages namely *chan-in-quoki Sacki* and *Sipie Setakha* having capacity of more than 1800 kg. These are wooden houses with galvanized-iron sheet roofs. In Mizoram, paddy storage structures are called *Buhwal Dahkhawn* which are circular in shape with capacity of around 2000 kg. The *Khasi* tribes of Meghalaya store their un-threshed paddy plants in a circular platform in such a way that all earheads are in the centre. *U-tihar* is circular bamboo structure used for storing paddy inside the house where as *Tyllu* is an outside wooden stack for storing paddy in large quantity. The *Adi* tribes of Arunachal construct a massive structure for paddy storage, which serves for 10 to 12 years.

Postharvest Processing of Horticultural Produce

The north-eastern region of India offers a favourable set of climatic conditions for cultivation of various types of horticultural crops such as fruits, vegetables, flowers, tuber and rhyzometic crops. A few crops like banana, citrus, pine apple, jackfruit, potato, cassava, sweet potato, ginger, turmeric, large cardamom, coconut, areca nut and a good number of vegetable crops are well adopted in the larger part of the region. Among ornamentals, orchids are unique for Sikkim, Arunachal Pradesh and Meghalaya having good commercial prospects. Certain other fruit crops like litchi in Tripura, kiwi and avocado in Sikkim, guava and low chilling peaches in Meghalya and spice crops like black pepper and cinnamon have good potential and deserve attention. Pine-apple, banana and citrus are the major fruits of NE region and these contribute about 75 per cent each in area and production. The total production of fruit crops in India is 33 million tonnes from an area of 3.2 million hectares in which NEH region shares 0.22 million hectare in area with annual production of 16.7 million tonnes.

Table 13.1: Popular Grain Storage Structure in Manipuri Tribe

Structure	*Shape*	*Size, m*	*Material*	*Capacity, tonne*	*Commodity*	*Type of storage*
Ningei	Square or rectangular	L = 3.0 B = 2.0 H = 2.0	Bamboo, plastered with mud and cow dung on inside wall	0.5-1.0	Paddy, maize, bean, pulses	Indoor
Kei	Rectangular	L = 3.0 to 5.0 B = 2.0 H = 2.0	Wooden plank or bamboo, mud	10-20	Paddy, maize, bean, pulses	Outdoor
Kot	Cylindrical	D = 2.5 to 3.0 H = 3.0	Bamboo plastered with mud and cow dung	1.0-2.0	Paddy, maize	Indoor
Apu Achouba (box)	Rectangular	L = 3.0 B = 2.0 H = 1.5	Wooden	0.5-1.0	Paddy, maize	Indoor
Chujak Mapun (maize bundle)	Hanging maize bundle in open hut	L = 5.0 B = 3.0 H = 0.4	Bamboo and thatch	0.05-0.1	Maize	Outdoor
Chujak yam (maize hut)	Thatched hut with walls	L = 4.0 B = 3.0 H = 3.0	Bamboo and thatch	1.0-2.0	Maize	Outdoor

Source: Annual Report, ICAR Research Complex for NEH Region, Umiam, Meghalaya.

While area and production of fruits and vegetables has been increasing at a fast pace, the postharvest losses are quite significant in the region due to improper cultivation practices, lack of elaborate harvesting machinery, non-availability of collection centres in major areas, un-suitable packaging containers, inadequate transportation system and poor processing facilities. Harvesting and postharvest operations of fruits and vegetables are still manual and handling, grading, transportation and storage has not yet received particular attention. The produce after harvest is hardly graded or packed before it goes to transport. Despite so much advancement in storage technology, the farmers and traders in this region are practising old storage methods and thus major portion of the produce is wasted during storage and transportation. Fruits infested by organisms are not attractive and require careful handling and packaging after harvesting. Pre-cooling treatment in fruits and vegetables is given to remove the latent or field heat, which reduces their respiration rate, minimize the susceptibility to micro-organisms and reduces water loss. Such treatments have not been initiated for prevention of postharvest losses of fruits and vegetables during transportation in warmer areas of north-east region. Waxing is done to reduce the evaporation loss of water from the fruits and thereby increasing the storability. No detailed survey has so far been conducted on the present practice of fruit collection, grading, sorting, packaging materials, storage methods etc.

Harvesting and Storage of Fruits and Vegetables

Most of the fruits are harvested when they are fully ripe, because growers are to sell their produce in local markets or even on the roadsides due to lack of proper marketing facilities. Oranges are normally harvested by hand-picking or by climbing on the tree directly or with the help of ladder or sometimes directly by beating the branches with a bamboo-stick. Pine-apple and banana are generally harvested by a sharp sickle-shaped hand tool locally called "Dao", whereas vegetables and tubers are harvested mainly with the help of spade. No technique has been standardized for harvesting of *Khasi* mandarin, indigenous varieties of banana and pine-apple in the region. In a systematic study, postharvest losses have been estimated as 12-19 per cent in banana, 8 to 30 per cent in orange and 5-20 per cent in pine-apple in India. The extent of losses of these fruits would be more severe in the north-eastern hill region, thus offering a challenging task to the researchers working in the region. The losses depend upon the fruit variety, storability as well as storage commodity. Unlike food grains, the perishable commodities lose their quality very quickly. Attention has to be paid for such perishable commodities right from harvesting to storage, transportation, processing and marketing. Indian Council of Agricultural Research (ICAR) Complex for NEH region, Umiam have developed harvesters, which can harvest fruits with petiole of 1 to 2 cm in length. This stops sap-bleeding, reduces latent infection and increases shelf-life of fruits by 2 to 4 days.

Storage of fruits and vegetables is another major problem in this region. Very often there is glut in the peak season and there is a wide disparity between growers' selling price and consumers' purchase price due to lack of proper storage facility. Adequate warehouse facility to store the surplus in case of increased production is lacking in almost every part of this region. However, the farmers practise their own method of storage of the surplus produce. For example, storage of orange and pine-apple is done in temporary huts in the vicinity of the house which are made of locally available material. In the huts, racks are covered from all the sides. Orange is kept on the racks in layer and covered with straw or plant-leaves to avoid exposure of fruits to the atmosphere whereas, pine-apple is kept without making layers. Thus the oranges and pine-apple are stored for about 20-30 days without any damage and farmers sell them according to their need and demand from the market.

Storage of Ginger

Ginger is the major rhizomatous crop grown in NEH region with annual production around 47,770 tonnes. In the traditional storage, rhizome is kept either *in-situ* or in pit storage. First of all, a pit is dug as per quantity of produce to be stored. At the bottom of the pit, layer of grass or paddy straw is laid and on which ginger or turmeric bulbs are kept horizontally in layers and this is again covered with grass or paddy straw. Thereafter the second layer of bulbs is kept in a similar fashion, and so on. Finally at the top, pit is covered and plastered with soil. While removing the bulb, only one side of the pit is opened and after taking out the required quantity, the opening is again filled with soil to avoid exposure of bulbs to the atmosphere. This method is followed in almost every part of the ginger and turmeric growing areas. Produce is kept fresh for about 4-5 months generally from November to February/March. This method is also being followed to keep seed of ginger and turmeric for the next planting season. However, in this method, more than 20-30 per cent loss is reported. In order to reduce losses and improve the efficiency of storage system, a storage structure has been developed by the ICAR Complex for NEH region, using locally available material *i.e.* bamboo mat, clay, sand and polythene sheet. The complete bamboo mat made of double walled structure is prepared with split bamboo structure. The cow-dung and clay mixture is filled in between the two walls and outer wall is plastered with mud. The rhizomes are stored in alternative layers of sand and rhizomes with sand at the bottom. Top of the structure is covered with polythene sheet. This structure can store about 4 tonnes of rhizomes. The cost of structure is around Rs. 2000.

Large Cardamom Drying

Large cardamom is an important cash crop for the Sikkim farmers and is gradually spreading to other north-eastern states. The quality of large cardamom depends totally on the external appearance and flavour. Appearance consists of colour, uniformity in size, shape and consistency whereas flavour consists of taste and odour, which are influenced by aromatic compounds. The method of drying plays an important role in this crop. The freshly harvested cardamoms are highly perishable and require drying upto 8 to 10 per cent moisture content so as to sustain it's the up-keep. The traditional drying kiln (*bhutie*) don't need the highly specialized skills and foreign input material. These kiln are made of locally available mud and stone masonry walls on all the four sides. An opening is provided in the front wall to feed the fire-wood. A bamboo mat is placed over these walls as a dry tray for holding the large cardamom. Height of the drying tray is normally 40 to 50 cm from the fire-place. These kiln have a temporary thatched roof. Such drying kilns are constructed by each cardamom grower near his farm. The raw cardamom is evenly spread over the drying platforms (mats) in thin layers of 15 to 20 cm. The fire-wood is burnt slowly in such a manner which generates lot of smoke. Hot smoke reaching the maximum temperature of 60°C passes through the cardamom layers and provides slow-drying. It takes nearly 35 to 40 hours of complete drying until capsules becomes hard and gritty in texture and turn dark-brown in colour. One must supervise the drying process for maintaining the optimum fire and stirring the capsules at regular intervals.

Performance of Zero Energy Cool Chamber

Spoilage of fruits and vegetables can be reduced by controlling the storage temperature and increasing relative humidity in and around the storage places. Refrigerated cold stores are not suitable for the rural areas due to high capital investment and energy requirements. Thus there is tremendous scope for adoption of small-capacity low-cost farm storage structures like Zero Energy Cool Chambers especially in the rural areas for short-duration storage of fruits and vegetables. Shelf-life of various fruits and vegetables like bitter gourd, capsicum, tomato, cauliflower, pine-apple and peach was evaluated inside and outside the cool chambers for two consecutive years. It was observed that the mean maximum temperature inside the cool chamber was about 5-6°C lower than the ambient one during summer season whereas during winter about 7-8°C temperature difference was observed. The relative humidity (RH) inside the cool chamber remains between 90-98 per cent throughout the year whereas outside in the open it varies between 67-75 per cent. Moreover, RH inside the cool chamber was nearly 17 per cent and 21 per cent higher during summer and winter months, respectively. The physiological loss in weight of each commodity was observed and recorded daily. It was observed that the shelf life of bitter gourd, capsicum, cauliflower and peach could be increased for 5 days and that of tomato and pine-apple for about 6 and 9 days, respectively, inside the Zero Energy Cool Chamber as compared to those under ambient conditions. Since the chamber costs only about Rs 2500/- and can be easily constructed in the rural areas, the fruit and vegetable growers can use it for short duration storage of horticultural produce.

Areca-nut Curing

Areca-nut *(Areca catechu)* is considered to be a native of Malaya and popularly known in India as *Supari* or *Shrivardhani.* In Meghalaya it is called as *Kwai* and is extensively grown in the region. There is around 6200 hectare area under areca nut cultivation, which yields around 4200 to 5900 tonnes of the produce every year. Areca nut is consumed in different manners in north-eastern states. The crudest way is to get the ripe fresh fruit, shell it with a knife cut into pieces and chew it with lime and betel leaf. In Meghalaya, the nut is mainly consumed after curing it under water. Curing is mainly done for taste, improvement in colour and easy removal of husk. The process helps in loosening the husk. This way husk can come out from the nut by hand; however still about 22 per cent of husk is removed with the help of knife. In Tripura state, the dry nut is consumed in small pieces with betel leaf and lime. No special type of curing is practiced in the state. In Assam, a circular pit is dug in the open area around the houses for curing nuts. The size of the pit varies from 2 to 3m in diameter and 3 to 4m in depth. The pit is plastered with leaves and cow-dung at the bottom, and inside the walls. The freshly harvested fruits are charged normally in April and early May for curing. The upper half-meter layer is again filled with leaves and cow-dung upto the ground level. The pits are opened in December-January for sale and consumption of the cured produce. Another popular method of processing of nuts in Assam consists of boiling the immature fruit in water for about two hours after the removal of husk and then drying in the sun. The produce is known as *Chikani.* The modern methods of nut processing consists of reducing the nut into slices, wafers or grits, followed by drying, flavouring, scenting, colouring and softening. Such processed nut is chewed with betel leaf and eaten sometimes directly. In Meghalaya, curing of areca nut is done in the flowing water. The technique might have been developed by trial and error method. Now the areca nuts are cured collectively as a community project in the Ri-War area of Meghalaya near Cherrapunjee.

Packaging

Fruits and vegetables being associated with high volume, weight, juice and tenderness require special packaging for transportation from one place to the other. Most of the conventional packings developed in this region on the basis of easy availability of local raw material are not suitable for the purpose. Survey conducted in local market of Shillong (Meghalaya) revealed that *Khasi* mandarins are normally packed in 20-25 kg capacity bamboo baskets while plums are packed in 5-10 kg wooden crates and bamboo baskets. Bamboo baskets are generally cylindrical in shape and fruits or vegetables are kept in different layers with paddy straw or plant leaves in between the layers, which works as cushion. Pine-apple after harvest is also usually carried in conical shaped bamboo baskets containing 20-25 fruits and carried manually to the roadside or collection centres. In such baskets and also in the trucks, fruits are arranged in crown-down position for transportation. The usual method of packaging of turmeric and turmeric-powder in the region is to pack them in jute bags without polyethylene lining, due to which volatile and colouring matter often get lost. Study conducted by the Central Food Technological Research Institute (CFTRI), Mysore revealed that aluminium-foil laminate gives maximum protection against loss of volatile oil, ingress of moisture and retention of colour. Thus for maintaining good quality of turmeric-powder during storage and long distance transport, recommended packaging material should be used. Since the north-eastern region is rich in forest vegetation and their products or by-products such as bamboo, cane, pine needle, sawdust are locally available, these can be utilized for preparation of cheap and cost-effective packages besides testing of improved packaging.

Plastic packaging materials have a lot of advantages over the others such as wood, glass and metal etc. In case of hilly regions in India, road communication systems are not well developed. Since the roads are highly sloppy, curved and uneven; food products during transportation get more damaged when packed in either wood, glass or metal containers as compared to plastic-packed foods. Plastic packaging materials are now-a-days becoming easily available in this region and there is an urgent need for replacement of other packaging materials by it. Bamboo baskets and gunny bags of different shapes and sizes are used for packing different fruits and vegetables, which are not very efficient and could easily be substituted by synthetic materials like polyethylene, nylon, polyester bags *etc.* However at some places packing cases, polyethylene bags and paperboard cartons are also being used for carrying various fruits and vegetables. Since in most parts of the region rainy season prevails quite long, there is need for large-scale introduction of plastic impregnated storage structures to creates awareness among farmers about various uses of plastics in day-to-day activities.

Use of Plastics in Seed Storage

Seed storage problem is considered to be the bottle-neck in developing a seed programme in the north-eastern region. Seed moisture and temperature are the main factors affecting seed storability. Hassington's thumb rules state that for every decrease of ten degrees Fahrenheit storage temperature, the storage life is doubled. North-eastern hill region is advantageous in having mild temperature which favours prolongation of seed storage life, provided the seed moisture is checked. Seed being hygroscopic in nature, responds fast to any increase in the atmospheric relative humidity even for a short spell of time. Seed moisture content could be checked with minimal cost through proper drying and packaging/storage in moisture proof containers/structures. Storage/packaging of seed with polythene of 700 gauge for being moisture-proof is recommended. Application of polythene in low cost seed storage structures needs to be explored under north-east conditions. The usual trend of seed usage in north-eastern hill region is the use of stored seeds for one off- season (from harvest till the next sowing *i.e.* for 8 months). This practice coupled with non-scientific production practices from generation to generation not only reduces the varietal purity but also risks the loss of variety in case of crop failure due to natural calamities. Storage of seeds for more than one sowing season *i.e.* 20 (8+12) months and more has been unsuccessful in the traditional way due to the humid and warm fluctuating temperature conditions. Thus, development of any low cost technique for seed storage would not only help in prolonging the varietal life span but also help to boost the agricultural productivity of this region through the use of superior quality seeds. Besides, with the varied micro-climatic areas of this region, adoption of a sustainable seed storage technology would greatly help in exploring the wide scope for seed industry particularly for the low volume, high cost seeds like those of vegetables and flowers. This would considerably contribute to the development of high profit seed industry for the small holding sized farmers. Studies have revealed that under the north-eastern hill ambient conditions, seeds could be stored normally for not longer than one off-season *i.e.* 8 months. Temperature and moisture not only lowers the seed physiological quality but also has direct impact on the micro-flora and insects. The prevailing conditions in the region thus favour rapid seed deterioration during storage. However, use of polythene containers could extend seed storability beyond two seasons. Further, use of charcoal or silica gel as desiccant inside the polythene containers exhibited a longer storability upto three seasons. A low cost seed storage bin model has been developed by the ICAR for small farmers. In this case, three materials are being used for making storage bin. The outer most layer is a thick bamboo-mat layer. The next one is a Hessian layer inside which contains the polythene (700 gauge) layer. The inner most layer is again a Hessian layer to protect polythene layer due to mechanical damage.

Packaging of Mushroom

Mushroom cultivation in the region is gradually picking up as cottage industry. Mushroom is a highly perishable commodity and generally spoils within a day. It is mostly sold fresh except some wild mushroom, which is also sold in dried form. For selling fresh mushrooms, open polythene bags with 250-500g capacity are used. Polypropylene bags are used for dried products. In a study at the ICAR Research Complex, Umiam, local wild mushroom *Letttinuia lateritia* was found to loose acceptability after 9 months of the storage in sealed polypropylene bags. In this region, no other information is available on packaging, storage and transport studies on mushrooms. Polystyrene or pulp board punnets may be used for packing large quantities of mushrooms to distant markets. The punnets are over-wrapped with differentially permeable PVC or polyacetate film. These over-wrappings create modified atmosphere in punnets producing an atmosphere having about 10 percent CO_2 and 2 percent O_2. It is expected that mushroom cultivation in the region would expand in future and plastics with proper disposal mechanism would go a long way in development of mushroom industry.

Transportation

Transport and communication system is absolutely poor in the entire region. Many areas, particularly the interior ones, are practically inaccessible in the hilly region. This creates problems of extension and supply of inputs besides prohibitive cost of transport outside the region causing disadvantages in the competitive market. The undulating topography, hill slopes and varied altitudes also create various problems for agriculture. Although roads have been built up during the recent years, yet the transport arrangements from growers' fields to the primary markets are not generally available. The produce is carried on head, back or shoulders by the producers and the average distance covered by them varies from 5 to 10 km. Produce is transported with the help of conical basket, keeping them on the back and supporting by hand from the bottom.

Transportation of Cuttings and Green-tea leaf

For transportation of tea cuttings for fairly long distance, the pruned stem are placed inside long polythene bags and arranged in boxes or baskets in such a manner that the leaves do not get bruised or broken. Packed in this way, the cuttings remain in good condition for about 24 hours even at an ambient temperature of 25-28°C. By further lowering the temperature by 10 to 20°C, the storage life is prolonged to 48 hours. For transporting over long distances by mail, pre-treatment of cuttings becomes necessary. Pre-treatment involves planting cutting in nursery beds at a closed spacing and lifting them after four to six weeks when a thin layer of callus tissue covers the cut surface at the base of the cutting. After lifting, cuttings are washed and the excess water is shaken off. These are then packed in polythene bags in bundles of 20-30 cuttings per bag. The sealed bags are packed in cardboard boxes for sending by mail. A large percentage of cuttings treated in this manner were observed to remain viable for 7 days. Introduction of tea-leaf carrying plastic bags made of HDPE polynylofab cum multifilament yams has brought marked change in the green-leaf management. These bags help in free air circulation and are hygienic and lightweight. A bag of the size of 20"x20"x 20" contains 10-15 kg green-leaf.

Food Processing and Agro-based Industries

The agro-industrial potential is vast in the north-eastern region. Potential of some crops *viz.* tea, jute, sugarcane, fruits like pine-apple, orange etc. as well as aromatic crops such as citronella have already been exploited. However the scope for agro-based industries from various other crops especially forest-based plants such as bamboo, canes etc. has not been fully utilized because of infrastructural shortcomings. The spices such as ginger, turmeric, black pepper, chillies, large cardamom and cinnamon can open avenues for agro-based enterprises. It is of interest to mention that quality of indigenous turmeric from Meghalaya has been found to be very high because of its higher *curcumin* content. Yams, sweet potato and maize tapioca are widely grown in the region. These plants are important sources of carbohydrate and starch and as such could be utilized for production of starch and as ingredient of livestock feed. Soybean grows extremely well in the entire region and can be used for producing various soy-products. There are around 30 fruit processing units in the region, processing primarily pine-apple and to some extent oranges and plums. Total installed capacity of over 15 public sector units is about 3000 tonnes but the utilization is only about 40 per cent of the installed capacity. Every year units suffer heavy losses and capacity utilization in public sectors is much below installed capacity. The cost of production is high due to the use of inefficient production processes, machines, excess man-power, poor management systems as well as high overhead costs. The cost of packaging material including cans, bottles, labels, cartons etc, is around 35 to 40 per cent of the variable production cost and 20 to 30 per cent of the total production cost as per survey conducted in the region. The cost of transportation to the markets outside the region also adds to the higher cost of the finished product. In view of remoteness of the area and poor transport facilities, postharvest technology like handling, storing, packaging and transportation need refinement to suit local conditions.

In Guwahati, certain units, process vegetables during the lean season. However, all the units suffer heavy losses every year mainly due to the absence of sales promotion, inefficient production processes, poor management systems, high over-head cost, non-utilization of by-product *etc.* The usual practice followed for the processing of turmeric in the Jaintia hills of Meghalaya, where turmeric is grown as commercial crop, is that, after harvesting, the bulbs are cleaned and cut into thin longitudinal slices and spread in the open for sun-drying. The dried slices are then powdered for marketing. Unlike in the plains where the turmeric fingers are usually boiled in water for sometime till the froth and white fumes with characteristic aroma of turmeric appears, no such practice is followed in hills. Pre-cooling facilities for fruits and vegetables are necessary to increase the shelf-life of fresh produce during transportation from production centre to the consumer centre. It helps in removing field heat, reducing the rate of respiration/transpiration and the amount of refrigeration required during storage and transport.

Soybean Processing

Soybean is a nutritious food having high percentage of protein and carbohydrate. The total area under soybean cultivation in Arunachal Pradesh, Meghalaya, Mizoram and Nagaland during year 1999–2000 was 17,400 hectares with total production of around 20,000 tonnes. People of Meghalaya consume soybean in the form of fermented product called *Tungrymbai.* The technologies used in the process of preparation of these food items are purely indigenous. *Tungrymbai* is prepared by the process of fermentation employing certain micro-organism. The preparation of *tungrymbai* consists of cleaning, soaking, removal of skin, cooking and packing and incubation. Cleaning soybean seeds is done to remove impurities like stone, mud, dirt, chaff, *etc.* followed by washing with water. After cleaning and washing, soybean seeds are soaked in water for about 8-10 hours. The outer skin is normally removed before cooking by rubbing the soaked seeds gently. After removing the skin, the soybean seeds are cooked with water for 5 to 6 hours. When it is well cooked, the product is allowed to cool down and is then packed. Packing and incubation is done only to convert the cooked soybean into the fermented one. This step is very important because it governs the quality of the ultimate product. For packing of the cooked soybean, special kind of leaves *(Clinogyne Dichotoma)* called *'Lamet'* (local name), are used. The cooked seeds are placed in a bamboo basket having its interior surface well lined with the *Lamet* and folded over the top surface of it. The top surface is also covered with thick cloth to provide insulation to the digester to maintain warm conditions. After proper covering, the digester (basket) containing the cooked soybean is kept over the cooking place in the house to maintain warm conditions to it. The incubation is done for 3-5 days to get the desired product. After the product is well fermented, it is ready to make different food items. The colour of this product is brown and the smell is somewhat unpleasant with mild pungency. However, since there was no additional cultured micro-organism added, it is difficult to explain the mechanism how the fermentation occurs. It is assumed that due to the wild fungi (or bacteria) present in the leaves used for packing and incubation, the fermentation occurs. The people of Meghalaya consume *tungrymbai* after making delicious food items with other ingredients such as green chilli, garlic, sesamum, slices of ginger, paste of mustard oil, salt, *etc.* For frying the fermented product, mustard oil is heated and paste of garlic, green chilli, turmeric powder and ground roasted sesame is put in it for deep frying after which the fermented soybean is added with salt to cook for about 20-30 minutes through continuous agitation. At the end of the process ginger slices are added and mixed properly and the product is then ready-to-eat.

Production of Dehydrated Ginger

Among the north-eastern states, ginger is grown in plenty in Meghalaya, Assam, Sikkim, Arunachal Pradesh and Mizoram. Besides, it is also cultivated in other states like Manipur and Nagaland. The traders procure ginger from the weekly market in different parts of the states, and transport these in truck loads to wholesalers at Dhubri, Guwahati, Calcutta and Delhi. The grower faces marketing problems due to dearth of local demand and are compelled to sell the surplus through middlemen at throwaway prices. Thus there is good scope to set up ginger dehydration units in north-eastern states as it will bring remunerative price to the growers and will cater to the needs of outside market also. Dry ginger is the main form in which ginger enters the international market. It is usually used in powdered form. Ginger based products have found multifarious uses in pharmaceutical brewery, soft drinks, perfumery, confectionery industry etc. Ginger oil and ginger oleo-resins are two other products that are in demand in the international market. They are used as flavours and possess certain advantages over the powdered spices. Exports of these products have also increased manifold in the recent past. In view of the increasing application of these products in the various food and allied industries in the country and abroad, there is a good scope for the development of ginger based industry in the region to meet the future requirements of this industry.

Tomato Products

Tomato is one of the most important vegetable crops in north-eastern states and widely grown in almost every state. Tomato is a valuable raw material used for processed products such as tomato juice, tomato puree, tomato paste, tomato ketchup etc. The recent scientific advances have revolutionized tomato-processing industries. Tomato-processing in north-eastern states is still not very significant. Recently, there was a steady rise in production and productivity of tomato due to introduction of high yielding varieties and sales promotion offered by the multinationals and creation of market infrastructure. With high fluctuations in market prices of fresh tomatoes in the urban market, there are good prospects for processing of tomatoes in the form of tomato ketchup, puree etc. in the household sector at the small scale level. Besides, the boom in the food service sector including fast food chain has widened demand potential for ketchup, soup etc. Experiments have shown that advertisement and publicity have influenced the pattern of consumption of tomato products. Besides, these products have good export potential especially in the Middle East.

Pine-apple Products

Pine-apple is one of the most important fruit crops in the north-eastern states. Total pine-apple production in north-eastern states during 1996-97 was more than 525 thousand metric tonne against the country's total production of 1071 thousand metric tonnes. Assam is the major producer followed by Meghalaya, Manipur, Tripura, Nagaland, Arunachal Pradesh and Mizoram. Bulk of the pine-apple production in India is consumed in fresh form. Proportion used for processing is less than 10 per cent in contrast to other pine-apple producing countries where more than 80 per cent of the production is consumed by the processing industry. There is only one large-scale fully automatic pine-apple processing unit in the country, rest of the units are only semi-automatic type. This shows the need for more pine-apple processing units. Due to lack of processing facility in the rural areas, pine-apples are sold fresh. However, processed pine-apple are liked by people more than the fresh ones. Thus there is a lot of scope for producing value-added products from pine-apple like squash, jam, jelly, juice, canned slices, nectar and other beverages with little investment. Pine-apple jam and slices in syrup are the most popular among these products because of improved flavour and taste. Cocktails, punches and salads are also prepared from pine-apple. Grated or crushed, it is used in puddings, frozen desserts, bakery and confectionery and ice cream etc. The core of the fruit is made into candies. Canned pine-apple products are also used for garnishing various sweet dishes.

Potential of Plant Based Chemical Industry

North-eastern India is blessed with nature's bounty as regards flora and fauna. Innumerable essential oil, medicinal plants, plants useful as dye, resin etc. occur in its forests and these have been in use by the tribes and others, in the region since time immemorial. Paradoxically, except for some saw-mill and plywood manufacturing units, there is hardly any forest based industry in the region, not to speak of more desirable photochemical industry at all, which should have been the premier industrial activity in the region long ago. In the post independence period, thanks to initiative taken by Regional Research Laboratory (RRL), Jorhat (Assam), Java citronella has become popular and has ushered in a new era in the field of photochemical industry in the north-east. Today, as much as 2400 ha of land are under the cover of citronella grass in the region producing about 900 tonnes of oil, which has given birth to a chain of citronella distillation units in the small scale sector. Reportedly, among all the essential oils produced in and exported from India, lemon grass oil used to rank second in value until few years ago. But the recent trend of indigenous production as well as export is dwindling. The traditional variety of lemon grass yields about 4-5 per cent of oil per kg of grass. Recently RRL Laboratory at Itanagar (Arunachal Pradesh) has successfully developed a new strain of lemon grass which yields 0.8–1.0 per cent of oil. Among other oil plants having good prospects in the region, mention may be made of Palma rose, Potchouli, Mentha arvensis, M. Piparita, M. Citrata, Vetiver and Turpentine. The fact that over 120 medicinal plants occur wild in the north-eastern forests shows the paramount importance of scientific studies and efforts for their commercial application. Apart from aromatic and medicinal plants, the north-eastern forests are also bountiful with quite a number of other plant species useful for production of vegetable tannins, dyes, edible oils, spices and condiments, intoxicants etc. Some selective items may be picked up for intensive scientific investigation and commercial exploitation for promotion of photochemical industry in the region.

The advantage offered by the photochemical industries is in the use of renewable sources of raw materials, low cost investment, high value products best suited for traditional agricultural communities as in the north-east ensuring great scope and prospect for development of industries in the field of drugs and pharmaceuticals, essential oils, food, agrochemicals, perfumery and cosmetic products etc. The rich flora of the north-east; its highly favourable agro- climatic conditions; almost total absence of any photochemical industry; growing standard of people of the region; hitherto virgin consumer market and increasing national and international demand for the products obtainable from the plant resources of the region- these hold out great hope and prospect for the promotion and growth of photochemical industry in the north-east with far reaching impact on its economy.

Marketing Constraints

Marketing infrastructure has not really been developed adequately in the north-east region even for field crops. One of the reasons may be that the commodities produced are not sufficient to meet the local needs and hence marketing was not considered an important aspect. However, this may be true to some extent for field crops. As for some horticultural crops, there are marketable surplus and for many other commodities domestic market within the region had also not been surveyed or developed. According to a study, by the National Productivity Council, 71000 ha area under cultivation of fruits in the region with an annual production of 630 thousand tonnes. According to this study, 95 per cent of the total ginger produced in the region, amounting to about 32,429 tonnes is marketable surplus. Similarly, 100 thousand tonnes of potato have been shown as marketable surplus. Absence of marketing and sales promotion activity, absence of sufficient domestic market and adequate storage facilities are some of the basic problems of marketing of horticultural produce. A detailed examination of the marketing cost in Ri-Bhoi district of Meghalaya reveals that 53 per cent of the marketing cost alone went to the various inter-mediaries as their margins and 24 per cent accounted for the transportation changes. The various inter-mediaries together appropriated a major portion of nearly 41 per cent of the consumer rupee of pine-apple in Meghalaya. Of the various inter-mediaries, the retailers retained the highest margin of 17 per cent of consumer rupee followed by agents/village traders and whole sellers who retained 14 and 10 per cent, respectively. Thus the pipe apple producers in Meghalaya are getting a meager share of consumer ruppee. In the remote areas of the hills, market is still at the rudimentary stage. Bangladesh imports apple from Bhutan and there should not be any reason why Arunachal apple (even after winning prizes in all India apple slow for excellent quality) cannot make a big headway in the markets of the region and other neighboring countries. In somewhat favourable situation, it has been reported that two villages of Mizoram sold orange worth 3.5 million rupees, three villages in Tripura for 5.4 million rupees and one village in Nagaland for 2 million rupees. North-eastern regional Agricultural

Marketing Limited (NERAMAC) was established in 1982 with the major objectives of promoting, organizing and executing development of production- procurement-processing marketing of horticultural products of north-eastern region, but it could not deliver the goods. Till date the corporation has incurred huge losses and now it has been referred to Board of Industrial and Financial Reconstruction under sick industries corporate act. The basic problems of marketing in the region that require attention are market density, enactment of market legislation, transport bottlenecks, storage facility including cold storage, market information/ intelligence, grading facilities and postharvest care etc.

As a result of the aforesaid discussion, following are some of the thrust areas identified in postharvest technology for value-addition and employment generation in the region:

1. Survey of existing postharvest practices such as harvesting, storage structure, packaging techniques, grading and sorting method and locally available products.
2. Refinement and adoption of the presently available postharvest technology for loss prevention and value-addition of agricultural products and other allied commodities.
3. Due to high rainfall, sunshine hours are not effective to dry the agricultural commodities. Therefore, devising the driers (including large cardamom drying) and storage structures to prevent the postharvest losses are the urgent needs.
4. Simple and efficient postharvest equipment for cleaning, grading, drying, value-addition, packaging and handling of important crops grown in the region should be introduced, demonstrated and popularized in the region.
5. Development of value-added products from different fruits and vegetables.
6. Introduction of processing equipment to produce value-added products on a small scale.
7. Evaluation of different pre-cooling techniques and standardization of manual and mechanical storage, waxing and grading processes.

8. Testing of packaging material and development of low-cost packaging structures using locally available materials for long distance transportation.
9. Introduction of agro-processing complex for a cluster of villages to generate employment and encourage the farmers to develop entrepreneurship.

References

Borthakur D.N. 1992. Agriculture of the North-eastern Region with special reference to Hill Agriculture. BEE Cee Prahashan Guahati.

Jain, Dilip, Satapathy, K.K. and Wahlang, E.L. Research Bulletin, Traditional Postharvest Technology of NEH Region, ICAR Research Complex for NEH Region, Umiam, Meghalaya.

Satapathy, K.K. and Agrawal, K.N., Ghadge S.V. and Jain Dilip. 2000. Bulletin on Research and Energy flow Pattem of a hilly village ecosystem in Meghalaya. ICAR Complex for NEH Region, Umiam, Meghalaya.

Singh, R.K.P., Satapathy, K.K. and Das, G. 2003. Presentation of fruits and vegetables. Research Bulletin, ICAR Research Complex for NEH Region, Umiam, Meghalaya.

Singh, R.K.P., Satapathy, K.K., and Sheo Govind. 1999. Present status of postharvest practices for fruits and vegetables in NEH Region. Agricultural Engineering Today, Vol (1-2), pp: 50-56.

Paul, B.K. 1991. Potential of plant based chemical industry in the north-east. Journal of the North-eastern Council 13 (3): 1-5.

Chapter 14

Nutri-Cereals: Value-addition of Coarse Cereals and Millets

R.C. Verma and S.K. Jain

Department of Processing and Food Engineering,
College of Technology and Engineering, Udaipur – 313 001

Small millets provide essential food for millions of people around the world. Most of the millets are grown as subsistence crops and therefore do not receive the international research and political support these deserve. Millets have certain specialties, which make these yield products of superior nutritional and technological characteristics than the major cereals. Millets have remained as the food for the people of lower socio-economic strata and traditional consumers because of their coarse texture, characteristic flavour, intense coloured seed-coat and cultural attachments. With constantly increasing awareness of good nutrition for healthy living, the consumption of millets is increasing among the affluent people also. However, the non-availability of processed products similar to rice or wheat is the primary reason for their consumption being confined to traditional consumers. There is a great need, especially in developing countries, to improve the productivity of the millets.

India is the largest producer of coarse cereals. Conventionally coarse cereals are considered to be all cereals except rice and wheat. The important coarse cereals generally referred are sorghum (*jowar*), barley, pearl millet (*bajra*) and maize. Finger (*Ragi*), *kodo* (*kodo*), foxtail (*kangani*), proso (*cheena*), barnyard (*sanwa*) and little millets (*kutki*) are a few other common types of millets. The nutrient composition of all these grains is comparable to rice and wheat. Some of the coarse cereals are even nutritionally superior. Mineral matter, especially calcium and phosphorus is high in most of the coarse cereals. Realizing their nutrient superiority, these grains are considered as *nutri-cereals* (Nutritious grains).

Being highly resistant to extreme environmental conditions, the millets are valued as potential crops which stretch the availability of food during seasons of mansoon failures or other agronomic stresses. Yet another important property most of the millets have is that due to their tough outer layers, insects find it difficult to bore through or cause damage and thus these can be stored without getting infested for years together.

The millets belong to the grass family. The grain size of the millets is very small ranging mostly between 2 to 6 mm. About 10 to 12 types of millets are grown in our country and the major among these are the pearl millet and finger millet (Table 14.1).

Table 14.1: Common, Colloquial and Botanical Names of Important Millets Grown In India

Common Name	*Colloquial Name*	*Botanical Name*
Sorghum	Jowar	*Sorghum bicolor*
Pearl millet	Bajra	*Pennisetum americanum*
Foxtail millet	Kangni	*Setaria italica*
Proso millet	Cheena	*Panicum miliaceum*
Barnyard millet	Sawan, Madira	*Echinochloa frumentacea*
Little millet	Kutki, Samai	*Panicum sumatrense*
Finger millet	Ragi	*Eleusine coracana*
Kodo millet	Kodo, Varagu	*Paspalum scrobiculatum*

Production

Small millets or minor millets are important feed crops for a large section of people in rural, tribal and hilly areas. Besides, the straw is often a precious fodder for cattle. Much of the production of these crops is by small landholders and farmers in the most marginal environments. Millets are crucial to the food security of many of the poor people in semi arid-tropics. Production of these grains has either remained stable over the past three decades or there is a tendency of slow decline. India is the largest producer of millet grains, producing about 33 to 37 per cent of the total about 28 million tonnes of the world produce. Small millets together occupy an area of around 4.5 million hectares, accounting for nearly 15-18 per cent of the area under coarse cereals. The annual production is 3.6 to 3.8 million tonnes accounting for nearly 18 per cent of the total production.

On account of being grown in rain-fed marginal lands with low or no agricultural inputs, the yield of these millets is very low in comparison to other cereals and ranges between 0.5 to 1 tonne per hectare.

Agro-economic Constraints

The cultivation of the coarse cereals and millets is mainly limited to poor and marginal soils with poor native soil fertility and rainfed cultivation with recurring drought. Negligible coverage under important varieties, no application of inputs like fertilizers and plant protection measures and the dearth of a remunerative market for produce because of the lack of appreciation about the nutritive value of these crops are the reasons that productivity of coarse cereals is less compared to other major crops. Coarse cereals remain the crops of difficult ecosystems with poor resource base, of poor farmers. Maize can however, be considered an exception to some extent. Although it is mainly a rainfed crop in much larger areas in Uttar Pradesh, Rajasthan, Madhya Pradesh and Bihar (during rainy season), it is largely an irrigated crop in the southern states.

The overall percentage of cropped area covered under irrigation is the least in case of coarse cereals. In this group too, maize has a slightly larger percentage area under irrigation (20.5 per cent), and in sorghum and pearl millet, the percentage area under irrigation is only 6.7 and 5.5 percent respectively. Among small millets, finger millet is most important with an annual acreage of 2 million hectares and production of 2.6 million tonnes. Often these crops are confined to hill slopes, undulating fields and poor soils, yet providing assured harvest at a reasonable level under marginal cultivation to resource poor farmers. It provides a broad idea of the importance accorded to coarse cereals in Indian agriculture.

This is one group of crops wherein available environmentally-benign technologies can bring about prosperity to the farmers through higher productivity with lesser and low-cost inputs and simultaneously ensures sustainability from fragile agro-ecosystems. Coarse cereals form an important component of dry land, tribal and hill agriculture. The poor economic base of the farmers with virtually no support from the Government for amelioration of their agro-economic environment is the reason that necessitates subsistence farming for sustenance of this deprived section of farming community. These crops at least ensure some production and simultaneously provide a fairly balanced nutrition.

Many tribal, backward and hilly areas where nutritious cereals and pseudo-cereals are grown are endowed with several physical, socio-economic and technological constraints. These include remoteness and inaccessibility of area, large physical and climatic variability, marginality of land, and people with low risk bearing capacity. The farm operations are mostly performed by women and available technologies are also not location-specific and many times not commensurate with the requirement of the area. The resources are not available. As such economic and ecological access to food continues to be a limitation that hinders the much needed food and nutritional security. Therefore, to deal with the situation in totality, there is need to address these areas for realising enhanced productivity and production of crops and commodities grown in such regions under varying farming systems/situations.

Pseudo-cereals

There are other grains, which are not exactly cereals, but are consumed the way cereals are consumed and are classified as "pseudo-cereals". Pseudo-cereals also constitute an important component of the life-support species and these crops have a high degree of acceptability in the poor farming households and sustainable agriculture although, at present these cover meager and scattered areas under cultivation. This group of under utilised crops includes amaranth (*Amaranthus spp.*), buckwheat (*Fagopyrum spp.*), chenopod (*Chenopodium spp.*) and Job's tears (*Coix lacrymajobi*).

Amaranth is a multipurpose crop with good potential for exploitation as grain, vegetable and fodder. The grains have high protein content with large amount of lysine and good balance of other essential amino acids. It is used as a staple food in the entire Himalayan region, southern India and to some extent in Gujarat, Maharashtra, Karnataka and eastern parts of Uttar Pradesh. Value-addition can be done to amaranth and many products such as baby cereal foods, candies and snacks, protein drinks and particularly hypo-allergic foods can be developed. Buckwheat is a short duration psedo-cereal especially suited for short summer seasons at higher altitudes. It thrives well on soils with poor drainage. Its flour is used during religious fasts by individuals/families and the plant is also used as pot herb with leaves as cooked vegetable. Chenopod is also an important vegetable-cum-grain crop for the hill regions and is generally mixed with other cereals and consumed. It is widely distributed/grown in patches in the Himalayan region. The pseudo-cereals are thus rich in energy and minerals and also possess important vitamins and amino acids in significant quantities.

Uses

All the coarse cereals besides being used as staple food by a larger section of people offer a good number of opportunities for industrial use as well. All these coarse cereals as well as millets can be utilized as human and animal feed also. Dr. Norman E. Borlogue stated that after the revolution in wheat and rice in the last two decades, next few decades will be known for maize. It is a truth that during the severe drought in Bihar during 1887, people harvested and ate green ears of maize-crop and survived under serious famine situation. Most coarse cereals can be puffed or popped. The pearling and polishing of millets can improve the appearance and reduce the fibre content. The flour of popped grain of millets provides a fine aroma to ready-to-eat products. The marketing of flour of coarse cereals is an important cottage industry. Refined flour is useful in the preparation of many snack foods. Especially, the extrusion cooking of millets appears to be highly promising in the preparation of value-added traditional and novel food products.

In hilly areas, four small millets *viz.* finger millet, barnyard millet, proso millet and foxtail millet have special status and form staple food of the poor particularly during the period of food scarcity. The foods prepared from the small millets have special significance as they provide warmth to the body during winter and also have special medicinal value and taste. Since ancient times, many traditional foods and drinks have been made from the proso millet and foxtail millet. Popped grains, especially of finger millet possess pleasant aroma and acceptable taste. Use of finger millet malt in low bulk, nutrient dense weaning foods has been well demonstrated and the technology is well adopted. Value-added products such as parboiled, flaked, popped, expanded, extruded-deep fried products and noodles can be prepared from proso millet. The starch of the grain is suitable as sizing agent in the textile industry.

The de-husked grain of little millet is cooked like rice and eaten or boiled in milk and served as sweet dish. Some times the grain is parboiled like rice. Often roti and porridge are made and consumed. It is also made into flour and used for making puddings and cakes. In parts of Northern India, de-husked grains are cooked in plain water and used either alone or in combination with finger millet or rice for preparation of Haria (fermented drink) or the grain is fried and eaten as *sattu*.

Finger-millet flour can be enriched with the addition of 5 to 10 per cent de-fatted soy-flour. Composite flour can also be prepared by mixing *ragi* flour with wheat flour/rice flour/sorghum flour. Special nutritional blends can also be made by mixing malted *ragi* flour with green gram flour to be used as weaning food.

Status of Millets

Millets are major subsistence food crops in many developing countries. These are more important than what statistics indicate because in the areas where they are grown for food, these are the only crops that people can grow as a source of energy and protein. Maize and wheat cannot replace millets as crops because millets can be grown even in extreme drought conditions of the semi arid tropics.

Small millets are traditionally the indispensable component of the dry land farming system. Their time of maturation is comparatively earlier than major cereal crops. They are ideally suited to diverse agro-climatic conditions of soil, water, temperature and humidity. Their cultivation stretches from the sea-level to almost 8000 feet above sea level. Under favourable climatic conditions more than one crop of these millets can be taken in one year. The grains are cooked like rice while in some areas the grains are ground to flour and used in the form of *Chapaties*. These also provide substantial quantities of palatable fodder for the cattle. Nutritionally, the grains of small millets are rich in protein, minerals and vitamins and are comparable or even superior to major cereals. The grains of small millets, unlike other cereals, are practically devoid of stored pests besides retaining viability for long periods even under poor storage conditions. All these have conferred a special distinction for small millets and appropriately referred to as famine reserves.

Small millets being eco-friendly crops demand very little application of chemical pesticides. These are the crops of fragile and vulnerable agro-ecosystems and should be considered as preferred species for a sustainable green agriculture. Small millets can be a stabilizing force in building up super cereals in industrial and food products. These crops thus can make significant contribution in natural resource management and ultimately lead to a more holistic approach in the maintenance of agro-diversity.

Place in Human Dietary

In spite of many advantages, millet grains hold only a secondary place in actual dietary consumption patterns and trends. As a class, they are considered coarse grains compared to rice and wheat. They are used as articles of human food only in the situations where other food grains cannot be raised or purchased at economic prices and they have largely remained the food of the poorer and less privileged sections of the population. In developed countries, where the production is quite large, they are used mostly as animal feed and for industrial purposes. Hardly 5 per cent of millets accounts for food uses and only refined and processed products derived from these have become popular. In developing countries, for compelling reasons of food economy, millets are used either as sole sources of grains or for supplementing rice or wheat. Even in these countries, the people who are accustomed to rice and wheat do not relish the millets. Some major features that mitigate their use are indicated below:

1. The outermost layers of bran/glumes are coarse and fibrous and if used as whole meal, are unpalatable. Being tough in nature make these irritating to the tongue and hence are not readily accepted by the people accustomed to wheat and rice.
2. These cannot be easily cooked like soft grains such as rice, as the sub-aleurone layers are hard and quite resistant to hydration thus rendering soft cooking in water difficult.
3. Millets take very long time to cook.

4. Their grains do not have the gluten characteristics of wheat and hence cannot be put to typical baking uses, hence preparation of thin *roti/chapatti* becomes difficult.
5. Most of these also have very strong characteristic flavour.

Processing of coarse grains other than maize for improving consumer acceptability has not received wide attention because these have remained the food of the poorer people of the society in developing countries. Further, the cost involved in processing is high enough to make the processed product beyond the reach of people.

Importance and Nutritional Quality of Millets

Small millets though occupy a relatively lower position among food crops in Indian agriculture; these are quite important from the point of food security at regional and farm level. The millets further contribute to the widening of food basket which at present is narrowing because of excessive dependence on a fewer food crops. The researches at the Central Food Technological Research Institute (CFTRI), Mysore, and the National Institute of Nutrition (NIN), Hyderabad, have found that the millets are superior to other major food grains including rice and wheat in many ways. All millets are distinctly superior to fine cereals in mineral composition and can contribute to effective nutritional security and human well-being.

Like other cereal grains, millets and pseudo-cereals also contain large proportion of carbohydrates about (65 per cent, mainly starch) and thus provide bulk of energy in diets based on these. A high proportion of this carbohydrate is in the form of non-starchy poly-saccharides and dietary fibre, which helps in the prevention of constipation, lowering of blood cholesterol and slow release of glucose to the blood stream during digestion.The composition of millets is shown in Table 14.2.

By virtue of their composition, some of the millets are even better with respect to the average protein content compared to rice. The pearl millet is rich in fat while the protein content of sorghum and pearl millet is comparable with wheat. Protein in millets ranges from 6-12.5 per cent and fat content from 1.1 to 5 per cent. *Ragi* has the highest calcium content among all the cereal grains (300-350 mg/100 g). It develops a high level of diastatic capacity upon germination. The capacity of elaboration of diastatic enzyme in *ragi* is next only to barley and similar to that of wheat. The essential amino-acid profile of the millets is also far better than that of wheat, sorghum or maize (Table 14.3). Millets have a well balanced amino acid profile and are high especially in methionine, cystine and lysine content.

Table 14.2: Comparative Analysis of Nutrient Composition of Cereals and Millets

Food Grain	*Nutrient Composition (per 100 g)*							
	Protein, g	*Carbo-hydrate, g*	*Fat, g*	*Crude Fibre, g*	*Miner-, als g*	*Cal-, cium mg*	*Phos-phorus, mg*	*Iron, mg*
				Cerearls				
Wheat	11.8	71.2	1.5	1.2	1.5	41	306	5.3
Rice	6.8	78.2	0.5	0.2	0.6	10	160	3.2
Barley	11.5	69.6	1.3	3.9	1.2	26	215	1.67
Maize	11.1	66.2	3.6	2.7	1.5	10	348	2.3
Sorghum	10.4	72.6	1.9	1.6	1.6	25	222	4.1
				Millets				
Pearl millet	11.6	67.5	5.0	1.2	2.3	42	296	74.9
Finger millet	7.3	72.0	1.3	3.6	2.7	344	283	3.9
Proso millet	12.5	70.4	1.1	2.2	1.9	14	206	10.0
Kodo millet	8.3	65.9	1.4	9.0	2.6	27	188	12.0
Foxtail millet	12.3	60.9	4.3	8.0	3.3	31	290	5.0
Barnyard millet	6.2	65.5	2.2	9.8	4.4	11	280	15.0
Little millet	7.7	67.0	4.7	7.6	1.5	17	220	6.0

These essential amino acids are of special benefit to those who depend on plant food for their protein nourishment. Millet grains are also rich in important vitamins *viz.* thiamine, riboflavin, folic acid or folate and niacin. It is quite interesting to note that lower incidences of cardiovascular diseases, duodenal ulcer and hyper glycaemia (diabetes) are reported among regular millet consumers. Millets are also rich in iron and phosphorus.

Most of the millets are generally used for the preparation of the *roti* (Indian pan cake), textured products (*badi*), sweet dishes (*sattu*) and porridge (*dalia*). Table 14.4 contains the nutritional value of

Table 14.3: Essential Amino-acids in Cereals and Millets (g/100g protein)

Food Grains	Value of Amino-acids (g/100g protein)										
	Isoleucine	*Leucine*	*Lysine*	*Methionine*	*Cystine*	*Phenyl-alanine*	*Tyrosine*	*Threonine*	*Trypto-phan*	*Valine*	*Histidine*
						Cereals					
Wheat	3.3	6.7	2.8	1.5	2.2	4.5	3.0	2.8	1.5	4.4	2.3
Rice	3.8	8.2	3.8	2.3	1.4	5.2	3.9	4.1	1.4	5.5	2.4
Barley	3.5	9.8	2.6	1.6	1.6	5.1	3.6	3.5	1.4	5.8	2.1
Maize	3.7	12.5	2.7	1.9	1.6	4.9	3.8	3.6	0.7	4.9	2.7
Sorghum	3.9	13.3	2.0	1.4	1.4	4.9	2.7	3.1	1.1	5.0	2.1
						Millets					
Pearl	4.1	9.6	3.4	2.5	1.8	4.8	3.3	3.1	2.0	5.5	2.5
Finger	4.4	9.5	2.9	3.1	2.2	5.2	3.6	3.8	1.6	6.6	2.2
Kodo	3.0	6.7	3.0	1.5	2.6	6.0	3.5	3.2	0.8	3.8	1.5
Proso	5.1	12.2	3.0	2.6	1.0	4.9	4.0	3.0	0.8	6.5	1.9
Foxtail	7.6	16.7	2.2	2.8	1.6	6.7	2.2	2.7	1.0	6.9	2.1
Barnyard	3.8	16.6	2.9	1.9	2.8	2.2	2.4	2.2	1.0	6.4	1.9

Table 14.4: Nutritional Value of Some Indian Dishes made from Small Millets (per 100 g)

Food Grains	Nutritional Value, per 100 g										
	Energy, Kcal	*Protein, g*	*Fat, g*	*Carbo-hydrate, g*	*Calcium g*	*Phosph-orus, g*	*Iron, g*	*Vitamin-A, IU*	*Thiamine, mg*	*Nicotinic acid, mg*	*Riboflavin, mg*
Ragi bade	133	2.4	2.3	25.8	0.10	0.09	1.8	68	0.15	0.36	0.04
Roti of *ragi*	249	4.3	4.9	47	0.22	0.16	3.2	137	0.27	0.70	0.07
Roti of maize	221	6.8	3.9	39.7	0.007	0.20	1.3	43	0.20	0.8	0.05
Roti of *Jawar*	168	5.0	0.87	35	0.01	0.13	3.0	60	0.16	0.87	0.05
Ragi puttu	289	3.3	5.1	57.7	0.14	0.14	5.5	193	0.14	0.4	0.04
Dalia of *ragi*	94	2.5	2.2	16	0.11	0.07	1.16	115	0.07	0.07	0.12

some of the common Indian dishes made from the small millets and various formulations developed for feeding programmes.

Traditional Processing Techniques

Traditional food processing methods play an important role in the utilization of locally available raw materials. It is unfortunate that not many attempts have been made to improve these existing traditional food-processing techniques or to improve the quality of indigenously processed foods.

Processing involves the partial separation and/or modification of the three major constituents of the cereal grain–the germ, the starch-containing endosperm and the protective pericarp. Various traditional methods of processing are still widely used, particularly in those parts of the semi-arid tropics where small millets are grown primarily for human consumption. Most traditional processing techniques are laborious, monotonous and carried out by hand. These are almost entirely left for women to do. To some extent, the methods that are used have been developed to make traditional foods to suit local tastes and are appropriate for these purposes. Traditional techniques that are commonly used include decorticating (usually by pounding followed by winnowing or sometimes sifting), malting, fermentation, roasting, flaking and grinding. These methods are mostly labour intensive and give a poor-quality product. Small millets would probably be more widely used if processing techniques are improved and if sufficient good-quality flour is made available to meet the demand.

In general, industrial methods of processing small millets are not as well developed as the methods used for processing wheat and rice. The potential for industrial processing of small millets is good. Custom milling may have a significant impact in our country like Nigeria, where about 80 per cent of sorghum and millets are now custom milled into whole flour.

In India many different kinds of traditional foods are made from small millet grains and they form the staple diet for many rural and urban households. *Ragi* is eaten in the form of *roti*. Many other traditional foods are made from popped *ragi* flour mixed with sugar/jaggery/ghee/milk/butter milk and salt. In several rural households a vast variety of traditional snacks are made from *ragi* and other small millets.

People in different countries prepare various products of finger millet in different ways. In India, the crop is usually ground into flour for cakes, puddings or porridge. A fermented drink or beer is made from the grain. The grain can also be malted. Flour from the malted grain makes a healthy food for babies and people who are sick or weak. In Africa, finger millet is used to make porridge, bread, malt and beer. It is also used for brewing. On adding a small amount of malted grain to a bowl of hot, starchy porridge, it turns into a watery liquid that can be fed to small or weak babies. If you eat a lot of starchy foods like wheat, rice, maize and potatoes *etc.* grain from finger millet helps your body to digest these foods. Some people use the straw from finger millet as food for working and milking animals. It is also used for thatching, for making walls for small granaries and for making dishes.

In times of distress or emergency when these coarse cereals are offered as part of the normal ration, people accustomed to rice or wheat find it very difficult to accept and utilize them as food. It is therefore essential to give some type of simple and unsophisticated processing to these grains to improve their acceptability. It must be emphasized that the refinement introduced during their processing should be moderate and not over-done so as to affect their nutritive value adversely.

Crushing or grinding to produce coarse grits or fine flour can reduce the particle size of the endosperm fraction. Women in their normal work atmosphere do this unpleasant hard work. Traditional milling stones used to grind whole or decorticated grain to flour usually consist of a small stone, which is held in the hand and a larger flat stone which is placed on the ground. Grain, which should be fairly dry, is crushed and pulverized by the backward and forward movement of the hand-held stone on the lower stone. The work is very laborious, and one can hardly grind more than 2 kg of flour in an hour. In a traditional process used, decorticated grain is crushed to coarse flour either with a pestle and mortar or between stones. Grain is also ground to coarse or fine flour in mechanized disk mills now located in many villages.

Flat breads are made by baking batters made with flour and water on a hot pan or griddle. Almost any flour can be used. The batter can be based on sorghum, millet or any other cereal. These flat breads are known by many local names like *roti* and *chapatti*.

Various traditional local crops produced in the southern part of Rajasthan are Barnyard millet (*batti*), finger millet (*mal*), foxtail millet (*kangni*), proso millet (*cheena*), barnyard millet (*kuri*), *kodra, hamlai, bhadla*. The tribal people consume these millets in various forms like *roti, dalia/thuli, kadhi, rabri*, etc. The tribes of southern Rajasthan also consume these millets as a substitute of rice after de-branning by pounding and boiling in water. *Jhajharia, rab, ghat, kheench, maize curry, ghugari, bati, missi roti, dhokle* and *papdi* are the various forms maize and sorghum, which are prepared and consumed in Rajasthan. Sorghum is occasionally popped.

Small Millets as Animal-poultry Feed

The small millet grain can become good substitute for other cereal grains in the preparation of concentrate mixture of live stock and poultry thereby reducing not only dependence on a few cereal crops but also the cost of feed. Hard pericarp and small seed size make them more resistant to insect damage and enhance their storage period. They contain about 7 to 12 per cent crude protein, 1 to 4 per cent fat, 70 to 85 per cent carbohydrate and 2 to 4 per cent minerals. Finger millet is low in fibre content (2 to 4 per cent) while other millets contain 7 to 10 per cent fibre. Proso millet is rich in amino acid lysine while the finger millet possesses high level of sulphur amino acid and thus these two millets appear to be promising in meeting the limiting amino-acid requirement of high yielding dairy cattle and sheep reared for high quality wool.

The small millet grains except finger millet have an outer husk, bran and starchy endosperm whereas in finger millet, seed coat is tightly fused with soft endosperm. However, from the chemical composition they appear to be equivalent to other cereal grains, which are commonly used in the ration of poultry, pigs and ruminants.

These crops can also be raised for fodder purpose and harvested at flowering stage. These yield good quality green fodder in a short duration of 40 to 80 days. The fodder is rich in crude protein (8 to 10 per cent), vitamins and minerals. These are highly palatable due to higher content of sugar and carbohydrates and form good material for making silage and hay. The straw/stovers of small millet form a source of dry roughage for ruminants. It contains 3 to 6 per cent crude protein and 30 to 40 per cent crude fibre. However, the digestibility of nutrients from these crop residues is generally low and is therefore, considered as poor quality roughage. There is a scope for selection of dual-purpose varieties which could give high grain yield as well as good quality straw. The threshed seeds of millets are coarsely grounded. The grounded millets, water and *Mahua's* flower are mixed in the ratio of 2.0:1.0:0.5. and given to the cows, buffaloes and oxen at the rate of 2 to 2.5 kg per day.

Storage of Millets

Storage of crops is an essential component of the whole production system. It facilitates to fulfil several farm objectives, *viz.* storing food for the future and avoiding food shortage, providing seed during the next growing season and allowing the farmer to sell the crop at a time when the price is good. The millets are grown primarily for own consumption by the farmers and for local trade. Therefore they are stored largely at domestic or household level.

Scientific attention to the storage of sorghum and millets has been considerably less than that for other cereals. The main reason is that sorghum and millets are regarded as minor grain crops despite their relative importance as staple food in many growing countries. The other notable reason is that farmers in the arid and semi-arid countries where millets are grown achieve quite impressive performance in grain storage by employing relatively simple traditional methods.

Most millets have excellent storage properties and can be kept for 4-5 years or even upto 30 years in simple storage facilities such as traditional granaries. This is because the seeds remain protected from the insect attack by the hard hull covering the endosperm and because grain is usually harvested and stored in dry weather conditions. Thus, despite large variations in production, year-by-year stocks are easily built up over the years.

Millet grains are known for good shelf-life. The grains dried to 10-12 per cent moisture can be stored for many years in the farmhouses. Proso millet (*cheena*) was found in a farm house of a tribe in southern part of Rajasthan stored for more than four decades without any adverse effect on grain quality and that too without any proper storage measures. The millet flour and their products also show good shelf-life. Besides India, in many countries of Eurasia including China and Japan, foxtail millet, proso millet and barnyard millet have been popular food grains for many centuries.

Millets may be stored, after drying and threshing, as loose grains in bags or loose containers. They are commonly left on the field, prior to threshing, in stacks or piles of harvested plants. The detached heads may also be stored away from the field, in exposed stacks or in traditional storage containers. However, the essential pre-requisites for storage of millets are the same as those for other grains.

Storage life of millets is inversely related to temperature and relative humidity during storage. Quality can be maintained by reducing storage temperature and humidity or moisture content (or all the three factors). Mould growth and intrinsic deterioration of millet in storage is negligible when the grains are sufficiently dry.

At rural and domestic levels, the types of storage structures used vary between regions. Bag storage is a common practice. Storing millets in jute bag facilitates aeration of the commodity but at the same time it favours cross infestation. Wooden crates dunnage is an ideal storage as it keeps the stacks 12 cm above the floor facilitating free circulation of air underneath. But these are seldom in use due to increase in cost and lack of knowledge. The traditional storage structures are largely unsuitable for fumigation. There are no chemicals generally accepted for the control of mould growth in stored coarse cereals. Small millets are usually stored in small quantities in traditional containers, often at the farms. Large quantities are seldom accumulated and bulk storage is rather uncommon.

Storage containers vary from small traditional on-farm or domestic containers to silos, which are sometimes found on large farms. Small granaries are made by weaving together various plant materials such as bamboo stalks, bark or even small branches and finally sealing in between gaps with mud or dung. These structures may be built directly on the ground or on the platforms so as to raise them from the ground.

Occasionally sorghum and millets are stored on the ground, usually un-threshed. The ear-heads are heaped in a pile (either indoors or outdoors) and covered with straw. As the grain is needed, ear-heads are removed and threshed. More often, grain is stored in gunny sacks, which are stacked either on the floor or on raised wooden platforms. Underground pits, which may be located underneath the house or outside, are also used. The pit is lined with paddy straw or sorghum straw. When it is full, the grain is covered with straw and soil. For longer-term storage, the top or roof of the pit is plastered with mud. Storage jars, silos and bins are made from a number of different materials. On the small scale, grain is stored in clay (earthen) pots. Larger containers are made from wood, brick or stone or from bamboo made into a basket, which is then sealed with clay or dung. When these containers are kept indoors they are sometimes left uncovered, but when they are kept outdoors they are covered with either a lid or a thatched roof. If the grain is to be stored for a long time, the top of the bin is plastered over with mud or dung. Occasional exposure to sunshine is the most commonly used measure for preventing insect infestation.

Flour is usually produced as it is needed and is not often stored for long periods because it tends to turn rancid. This is particularly evident in case of pearl millet flour, which has a very high fat content. Sorghum and millets, particularly pearl millet, are therefore best stored as whole grain.

Scope of Processing

The first objective of processing is usually to remove some of the hull or bran–the fibrous outer layers of the grain. Pounding followed by winnowing or sieving usually does this. The grain may first be moistened with about 10 per cent water or may be soaked overnight. When hard grains are pounded, the endosperm remains relatively intact and can be separated from the heavy grits by winnowing. With soft grains, the endosperm breaks into small particles and winnowing and screening can separate the pericarp.

Finger millet is a staple crop in many communities. It is a source of good, healthy food during difficult times, such as drought or crop failure. It's relatively cheaper, easy to grow and can be stored well for years. There are many other good reasons to grow finger millet. A single seed produces a lot of grains, so the seed is not expensive. It doesn't need any extra fertilizer or water to grow well. Finger millet can be grown almost anywhere, including hilly areas and can be planted with other crops. It can be stored well for long periods of time and that's why finger millet is often called a famine crop. Finger millet stores well because it has small seeds. This means that the seeds dry quickly. Insects can't get inside the grains to damage these during storage. Finger millet generally does not have problem of pests. Even in a humid area, the millet seeds can be stored much longer than the seeds of cereals. It doesn't rot when it's stored during the wet months. Finger millet is also a nutritious food. It contains many essential nutrients that people need to stay healthy. One of the best things about finger millet is that various nutritious food products can be cooked quickly. The small millet seeds take less time to cook.

When suitably prepared grain is pounded, the bran fraction contains most of the pericarp, alongwith some germ and endosperm. This fraction is usually fed to domestic animals. The other fraction, containing most of the endosperm and much of the germ along with some pericarp, is retained for human consumption. Retaining the germ in the flour improves its nutritional quality, but at the same time increases the rate at which the flour becomes rancid. This is particularly important in case of pearl millet.

Dry, moistened or wet grain is normally pounded with a wooden pestle in a wooden or stone mortar. Moistening the grain by adding about 10 per cent water facilitates not only the removal of the fibrous bran, but also the separation of the germ and the endosperm, if desired. Although this practice produces slightly moist flour, many people temper the grain in this way before they pound it. Pounding moist or dry grain by hand is very laborious, time consuming and inefficient. A woman can decorticate hardly 1.5 kg of the grains per hour with pestle and mortar. Pounding gives a non-uniform product that has poor keeping quality.

The millet grains offer many opportunities for diversified utilization and for value-addition. With proper processing it is possible to make many different kinds of food products by adopting appropriate milling, popping and other technologies. Except finger millet, other millets resemble rice in grain morphology containing husk, bran and endosperm. Traditionally the husk and bran are separated by hand pounding. However, in recent years milling technology has been improved to enhance the grain quality and to save time, as well as energy. Millet mill is available today both for cottage level and large scale processing. Milled millet can be further processed towards various food uses such as flakes, quick cooking cereals, ready-to-eat snacks, supplementary foods, extrusion cooking, malt based products, weaning foods and more importantly health foods.

Finger millet flour is easy to make since the endosperm and bran are pulverized freely and in such a flour fibre content is normally higher. However, it is possible to reduce fibre content by adopting simple sieving methods. Malting of *ragi* for food uses has been in practice from the times immemorial. *Ragi* has superior malting properties compared to other cereal grains like rice, maize, *jawar* and *bajra*. *Ragi* contains high level of calcium and its protein is rich in methionine and sulphur containing amino-acids. While malting finger millet does not pose problems of mould growth, off odour etc., finger millet malt has acceptable taste, very good aroma and a longer shelf-life.

Extrusion is being used increasingly for the manufacture of snack foods. In extrusion processes, cereals are cooked at high temperature for a short time. Starch is gelatinized and protein is denatured which improves their digestibility. Anti-nutritional factors get inactivated. Micro-organisms are largely destroyed and the product's shelf-life is thereby extended. The products can easily be fortified with additives.

Alternative Uses of Sorghum and Millet

These grains can be used for traditional as well as novel foods. However, there is a need to look into the possibilities of alternative uses. Though sorghum and millets have good potential for industrial uses, they have to compete with wheat, rice and maize. Sorghum in particular could be in great demand in the future if the technologies for specific industrial uses are developed. Although pearl millet has some potential for industrial use, other millets have limited potential because of their small grain size and the associated difficulties of adopting a suitable dehulling technology. However, these can be used for animal and poultry feed. There is a need to compare their performance as feed with maize.

Sorghum and millets can be adopted for other food products by using appropriate processing methods. It may be possible to select grain types with improved milling quality that will make these crops competitive with other cereals in terms of utilization. Wheat milling technology with suitable modification can be effectively used for grinding sorghum and millets. Although bread can be produced from whole sorghum flour, the quality of the bread can be improved by using sorghum flour from which the bran fraction has been removed by passage through sieves. Sorghum malt can be used to make biscuits, weaning foods and beer. Addition of sorghum malt up to 40 per cent in biscuits causes reduction in stack height and increase in spread because of increased water absorption.

The use of sorghum in common foods such as *idli* (a steamed product), *dosa* (a leavened product) end *ponganum* (a shallow-fat-fried product) can be popularized for wider use in sorghum-growing areas. A few important sun-dried or extruded products from sorghum are *papad, badi* and *kurdigai*. These products usually have a shelf-life of over one year. These can be popularized through marketing channels similar to those used for rice products. To conclude, efforts should be made to market these foods commercially to popularize these more and more amongst a greater population.

References

Rai, Mangla and Mauria S. 1999. Coarse cereal: Not for the poor alone. The Hindu Survey of Indian Agriculture. pp. 55-58.

Rai, Mangla. 2002. Amazing nutrient composition. The Hindu Survey of Indian Agriculture. pp. 59-62. Seetharama N. and Rao B D (2004) Sustaining nutritional security. The Hindu Survey of Indian Agriculture. pp. 37-38.

Seetharam A. 1997. Coarse cereals: Waiting for incentives. The Hindu Survey of Indian Agriculture. pp. 55-59.

Seetharam A., Kadalli, G G and Halaswamy B H. 2001. Results of front line demonstrations and Technology for increasing Production of finger millet and small millets in India. All India Coordinated Small Millets Improvement Project, University of Agrl. Sciences, GKVK, Bangalore 560 065.

Chapter 15

Postharvest Handling and Management of Horticultural Crops in North-eastern Region

D.S. Yadav and R.K. Yadav

Division of Horticulture, ICAR Research Complex for NEH Region, Umiam – 793 103, Meghalaya

The north-eastern hills (NEH) region of India comprising the states of Arunachal Pradesh, Assam, Manipur, Meghalaya, Mizoram, Nagaland, Sikkim and Tripura (21.50°-29.50° N latitude and 85.5°-97.5° E longitude) represents a distinct agro-climatic zone of the country. Morphologically NEH region of the country is marked by the development of a series of ridges and valleys, terraces, several geomorphologic or planer surfaces at different elevations up to 1,500 m above the mean sea level. The mean annual rainfall in the region exceeds 2000 mm. The region, though having around 8 per cent of total geographical area of the country, receives 12.1 per cent of the total precipitation. The soils are usually rich in organic matter and acidic to strongly acidic (pH 4.5-5.0) in reaction

The total area under horticultural crops is around 822.5 thousand hectares, which is around 3.14 per cent of the total geographical area of the region (Agricultural Research Data Book, ICAR-2002). The area under various fruit crops in north-eastern region is 270.4 thousand hectares and production is 2337.7 thousand tonnes with average productivity of 8.65 tonnes per hectare. The total area under vegetable crops in the NE region is 367.9 thousand hectares and production is 4051.8 thousand tonnes with productivity of 11.01 tonnes per hectare. The diverse agro climatic conditions, varied soil type and abundance of rainfall offers immense scope for cultivation of different types of horticultural crops, including fruits, vegetables, flowers, plantation crops, tuber and rhizomatous crops and crops of medicinal and aromatic importance. The fruits grown in this region, range from tropical and sub-tropical like banana, papaya, pine apple, jack fruit and citrus to temperate fruits like apple, pear, peach, plum and even certain nut fruits. The region has rich diversity of different vegetable crops and both tropical and temperate vegetables are grown to a considerable extent. The major vegetables grown in the region are brinjal, cabbage, cauliflower, *okra*, onion, pea, potato, tomato, *knol-khol*, radish, carrot, French bean and different cucurbitaceous crops. Among the flowering plants special mention may be made about the orchids, where about 600 species are reported to occur in the region alone. The other commercial flowers of the region are marigold, tuberose, gladiolus and chrysanthemum. Tuber and rhizomatous crops like tapioca (cassava), sweet potato, *Dioscorea*, colocasia, ginger and turmeric grow abundantly in the region, while plantation crops like tea, coconut, areca nut, cashew nut have considerable impact on the economy of the tropical and sub-tropical parts of the region. Other plantation crops like rubber and coffee, medicinal and aromatic oil yielding plants like *Solanum spp., Dioscorea spp., Cymbopogon spp., Citrunella spp.* etc. have also been considered suitable for certain areas of the region.

The region has tremendous potential for horticulture development but the high rainfall (3000 to 5000 mm/year) and high humidity persisting through out the year is highly congenial for growth and development of micro-organisms. The postharvest losses vary from 5 to 40 per cent depending upon the crop and variety. The postharvest losses in case of banana have been recorded from 15-20 per cent, where as in case of pine apple, *Khasi mandarin* and tomato are 12-17, 21-25 and 21-30 per cent respectively. The postharvest losses in perishable commodities are higher in comparison to other parts of the country due to lack of mechanization, collection centers in major producing areas, suitable container, commercial storage houses, poor transportation, unorganized marketing system and processing units. Under such a situation, besides increasing production, the reduction in postharvest losses will be a complementary means for increasing supply of fruits and vegetables. This approach has an added advantage because the cost of preventing postharvest losses is less than producing a similar additional amount of the same quality. Therefore, attention has to be paid for such perishable commodities right from harvesting to processing and marketing.

Prevailing Practices of Postharvest Handling and their Management

Maturity Standards

Correct stage of maturity of fruit is very important at harvesting time to reduce postharvest losses. This maturity is indicated by several parameters based on visual, physical and chemical indices. These not only help in deciding the date of harvest but also to sort out the harvested fruits into lots of varying maturity for their optimum use. Maturity standards for mango, guava, sweet oranges, grape, litchi and *ber* have been worked out. The Bureau of Indian Standards has suggested guidelines for maturity standards in banana. For banana, judgment of degree of maturity is based on the number of days from flower emergence, pulp/feel ratio, colour of pulp and odour. Central fruit of normal shape from the outside row is sampled. Banana fruit bunches mature over a period of 90 to 150 days after the inflorescence emerges from the pseudo stem. Pine apple can be harvested at maturity within 150 to 160 days after flowering. Studies conducted at ICAR Research Complex for NEH region, Umiam shows that Khasi mandarin harvested during November can successfully be stored for 45 days.

Harvesting

Most of the fruits and vegetables are harvested when they are fully ripe, because the growers sell their produce in local markets or even on the road side due to lack of proper marketing facilities. In north-eastern region of India, the harvesting of fruits is still done manually. Oranges are normally harvested by hand-picking or by climbing on the trees directly or with the help of ladder or sometimes directly by beating the branches with long bamboo stick. Pine apple and banana are generally harvested by a sharp sickle-shaped hand tool locally known as 'Dao' whereas tuber and rhizomatous crops are harvested mainly with the help of spade. The vegetables like tomato, capsicum, cucurbits, legume and cole crops are harvested manually. The techniques for harvesting of banana, citrus, grape, papaya and pine apple are available elsewhere in the country. But no such techniques have been standardized for harvesting of *Khasi mandarin*, indigenous varieties of banana and pine apple in this region. The most of the horticultural produce like citrus, banana, pine apple, tomato, capsicum and bean are collected in bamboo baskets locally called 'Thapas' while harvesting or plucking, therefore produce does not come in contact with soil and its quality is better.

In general, manual harvesting of fruits is followed in India. However, different fruits require different methods of harvesting. Harvesting of fruits and vegetables should be carried out during the cooler part of the day *i.e.*, in the early morning to avoid higher levels of field heat. Picking and aching methods and hand-harvesting cause less damage than mechanical methods. Moreover, the Indian Institute of Horticultural Research, Bangalore, and Central Institutes like ICAR Research Complex for NEH Region, Barapani, have developed a harvester, which can harvest fruits with pedicel of 1 to 2 cm. in length. This stops sap bleeding and reduces latent infection and increases shelf-life of fruits by 2 to 4 days.

Pre-cooling

Pre-cooling facilities for fruits and vegetables are necessary to increase the shelf-life of fresh produce during transportation from production centre to the consumer centre. It helps in removing field heat, reduces the rate of respiration and the amount of refrigeration required during storage and transport. Methods of pre-cooling are air-cooling, hydro-cooling, vacuum cooling, evaporative cooling and contact-ice cooling. But such facilities are non-existent in this region.

Grading

The need and advantages of grading are well accepted. Though at central level, Agmark grades and grade standards based on the variety, appearance, size, colour etc. have been fixed for 17 fruits and vegetables, but grading practices have not been prevalent in this region due to lack of facilities in the existing market. However, certain fruits and vegetables like pine apple, orange, guava, banana, cucurbit, tuber crops, tomato and capsicum are sold as per their size in different grades. However, grading of fruits and vegetables is not common at every place and it is done only at selected places in the cities. In case of banana and apple, market grading is made on varietal basis. In case of large cardamom, which is grown at commercial scale in Sikkim, capsule quality is mainly accounted for its external appearance, flavour and nature or by-products. The grading is done through visual perception of colour, uniformity in size, shape, consistency and texture.

Establishment of grading/sorting centers at the farm level is essential to reduce the postharvest losses. Grading includes removal of split, punctured and deformed fruits. It will also avoid spoilage of good produce which may otherwise come in contact with damaged and diseased produce. This can be done manually or mechanically.

Packaging

The function of a package is primarily to contain and protect the produce. The size of the package is therefore important and should be designed in terms of the amount, the market or customer requirement in a single unit. In other circumstances the package size may be dictated by what a person can reasonably lift or carry. Fruits and vegetables being associated with high-volume, weight, juice and tender by nature require special care like good packing for transportation from one place to other. Most of the conventional packing developed in this region on the basis of easy availability of local raw material is not found much suitable for this purpose. For example, bamboo baskets and gunny bags of different shapes and sizes are used for different fruits and vegetables, which is not very efficient and could easily be substituted by synthetic material like polyethylene, nylon, polyester bags etc. Now at some places packing cases, polyethylene bags and paperboard cartoons are being used for carrying various fruits and vegetables.

In local markets mandarins are normally packed in 20-25 kg capacity bamboo baskets while plums are packed in 5-10 kg bamboo basket. Bamboo baskets are generally square, cylindrical or conical in shape and fruits or vegetables are kept in different layers with paddy- straw or plant-leaves at the bottom and top, which provides as cushioning material. These fruits are re-packed in wooden boxes, which contain 5 and 10 kg fruits for sending outside the region. The shelf-life of *Khasi mandarin* fruit increased upto 45 days when packed in polysheet (50 gauge) and stored at room temperature (20 ± 2 °C). Similarly in case of corrugated fibre board (5 ply) the shelf-life increased upto 21 days by using newspaper cuttings as cushioning material. Pine apple is also usually carried in conical or square/ rectangular shaped bamboo baskets containing 20-25 fruits and carried manually to the roadside or collection center. Peach and plum are brought from the farmers' fields in bamboo baskets and repacked in small cartons of 2-5 kg. Due to the absence of suitable packaging material and being highly perishable, plums and peaches are not transported to distant markets. The vegetables like tomato and capsicum are also packed in bamboo baskets (Thapa) after properly placing paper or banana leaves on each side to protect bruising. The cabbage, carrot, radish and other fresh vegetables are packed in bamboo baskets. However, the cauliflower curds are individually covered with newspaper and packed in ventilated bamboo baskets. The *chow-chow* (*chayote*) and sweet potato are generally packed in gunny bags of 25–40 kg for sending it to the collection centers. ICAR Research Complex, Barapani has standardized different packaging materials for important vegetables. The capsicum and brinjal were stored for 16 and 18 days respectively in perforated PET (polyethylene Terephthalate) film of 150 gauge.

Ginger and turmeric are packed in jute bags without polyethylene lining due to which volatile and colouring matter often gets lost. However, the turmeric powder (processed at home scale) is packed in polysheet. Study conducted by Central Food Technological Research Institute, Mysore revealed that aluminum-foil laminate gave the maximum protection against loss of volatile oil, ingress of moisture and retention of colour. Thus for maintaining good quality of ginger and turmeric during storage and long distance transport such recommended packaging material should also be used in this region.

Studies have indicated that the shelf-life of fruits and vegetables could be doubled under normal conditions of storage by pre-packaging them in appropriate packages with adequate ventilation by controlling physiological changes and mechanical damage. Since NEH region is rich in forest vegetation and their products or by-products such as bamboo, cane, pine needle, sawdust etc which are locally available, these can be utilized for preparation of cheap, cost-effective and eco-friendly packaging material. Proper containers for transportation and marketing of fruits and vegetables can greatly reduce their losses and maintain their freshness and quality for a long time. Broadly, package is of two types *viz.* Consumer pack and Transport/market pack. Both of these should be designed in such a way that there should be minimum loss to quality during handling and transportation. The demand of the future would be for a package, which could keep the product as fresh as possible till the time of its consumption. The package may be very attractive, light in weight, unbreakable under normal conditions, easy to handle, disposable and which will not change the odour, taste or flavour of the product.

Transportation

Transportation is a serious bottleneck in this region due to non-existence of adequate road and railway lines to connect the growing centers with the primary and secondary market. Horticultural crops generally being located on hills and in remote areas, transportation of harvested crop to collection centers or storage places is done manually. Produce is transported with the help of conical baskets (locally called Thapa), keeping them on the back with the help of rope-pad on the head and providing support by hand from the bottom. Farmers carry their produce through head-load for long distances ranging from 1-15 km on hilly slopes. Where roads are available, farmers bring their produce to the nearest market by loading it in the local bus or jeep after packing these in bamboo baskets or gunny bags without grading. The road condition is very poor at many places due to which the produce gets damaged during transportation and deterioration starts very fast. There are still many places in the region where roads are not available and these are not connected with the market. Transportation contributes the most (about 15 per cent) to the total postharvest losses. It varies from 6-31 per cent depending upon the crop and varieties. Thus it is a vital stage during which maximum care is to be taken so as to minimize the damage to fruits and vegetables from bruising, vibration, shock load, impact load and bulk load etc.

Storage

Storage facilities for perishable commodities such as fruits, vegetables and spices are important for avoiding loss due to spoilage, to preserve the quality of the produce and to ensure a fair price by making the produce available for a longer period, thus avoiding the glut in the market during the peak period of harvesting. However, storage facilities are non-existent for such perishable commodities. Due to lack of storage facilities of fruits and vegetables very often there is glut in the peak season and there is a wide disparity between growers' selling price and consumers' purchase price. Adequate warehouse facility to store the surplus in case of increased production is lacking in almost every part of this region. However, farmers practice their own method of storage of the surplus produce. For example, storage of orange and pine apple is done in temporary huts prepared near the house made of locally available material. In the hut, racks are made and it is covered with straw or plant leaves to avoid exposure of fruits to the atmosphere, where as pine apple is kept without making layers. By this method oranges and pine apples are stored for about 20-30 days without any damage and farmers sell them according to their need and demand from the market.

On the other hand, storage of ginger and turmeric is done in the dug out pits. First of all a pit is dug as per the quantity of produce to be stored. At the bottom of the pit, a layer of grass or paddy-straw is laid and on this ginger or turmeric is kept horizontally in different layers. Finally at the top, the pit is covered and plastered with soil. While removing the rhizome, only one side of the pit is opened and after taking out the required quantity, the opening is filled with soil to avoid exposure of rhizomes to the atmosphere. This method is followed in almost every part of the ginger and turmeric growing areas. Produce is kept fresh for about 4-5 months, generally from November to February-March. This method is also being followed to keep seeds of ginger and turmeric for the next planting season. In certain areas like Arunachal Pradesh, farmers prepare a small bamboo/wooden house about one meter above the ground. Straw or dried leaves are spread at the bottom. Above the straw the ginger/ turmeric rhizomes are put in layers and finally covered with straw. This system of storage avoids the damages caused by rats etc. The tuber crops like potato, colocasia and sweet potato are generally stored in the huts after spreading a thick layer of gunny bags, straw or tree leaves. However for seed purposes the tuber crops except potato are stored in the soil after making a small pit and when these start sprouting are dug-out and used for sowing after removing the damaged ones. There are only a few cold storages in the region and these are generally used for storage of potato only. For other horticultural crops no such system is available.

Storage at low temperature immediately after harvesting reduces the rate of respiration resulting in reduction of build-up of the respiration heat, thermal decomposition, microbial spoilage and also helps in retention of quality and freshness for longer periods. The storage life is governed by several factors *e.g.* variety, stage of maturity, rate of cooling, storage temperature, relative humidity, packaging system etc. The optimum refrigerated storage requirements for different fruits have been worked out as 1.7-3.3° C for apple, 12.8° C for banana, 0-1.7° C for grapes, 5.5-7.2° C for orange and 8.3-10° C for guava, pine apple and mango. Based on the principle of evaporative cooling, low-cost, low-energy cool chambers have also been developed for storage of fruits and vegetables, which can reduce the storage temperature by 17-18°C. Controlled or modified atmosphere (CA/MA) is also an important innovation in storage of fruits and vegetables. In this the produce is stored in an atmosphere with more CO_2 and N_2 but less O_2 than in the normal air. These methods considerably retard the respiratory activity of the stored produce.

Low-pressure storage is another method for storing fresh fruits under partial vacuum. In this type of storage, chamber is continuously vented with water-saturated air to maintain the O_2 level and minimize water-loss.

Temperature, relative humidity and air velocity are the major factors which have considerable effect on storage behaviour of fruits and vegetables. Therefore, sufficient care should be taken for these parameters to make their combination most favourable for enhancing the life of perishables. The NEH region also requires such improved facilities for storage.

Processing

In the region most of the fruits and vegetables are sold afresh and less than one percent of the total production is being processed for other value-added products like jam, jelly, squash, sauce, pickle, wine etc. In marketing of fruits and vegetables particularly in this region where sufficient quantity of marketable surplus is available due to transportation and marketing difficulties, processing of fruits and vegetables may be an alternative sector through which a reasonable return can be ensured to the growers. There are about 31 processing units spread throughout the region. Most of these units are processing pine apple and citrus with negligible amount of guava and other fruits. Certain units like M/s Fruitos and Company, Guwahati, process some vegetables in lean season. However, all the units suffer heavy losses every year mainly due to the absence of sales promotion, sales organized at individual level, inefficient production process, poor management system, high over-head cost, non-utilization of by-products etc. The usual practice followed for the processing of turmeric in the Jaintia hills of Meghalaya, where turmeric is grown as commercial crop, is that after harvesting the bulbs are cleaned and cut into thin longitudinal slices and spread in the open for sun-drying. The dried slices are thereafter then powdered for marketing. Unlike the plains where the turmeric fingers are usually boiled in water for sometime till the white froth and characteristic aroma of turmeric appears, in this region no such practice is followed.

The processing industry for extraction of oil from fresh ginger has been established recently in two states *i.e.* Manipur and Mehalaya to process the surplus ginger from this region. In Sikkim the fresh large cardamom capsules, which are highly perishable and contain moisture upto 85 per cent are immediately cured to retain only 10-13 per cent moisture on dry weight basis. The retention of natural colour and flavour during curing process is the most important. The flavour constituents are highly volatile and easily lost due to direct heat and higher temperature (>55 °C). However, the packed hot humid condition of curing chamber may result in discoloration and oozing of capsules. Earlier the curing was done only by Kiln system in which a thin structure of bamboo just like a mesh was put over the kiln on which freshly harvested fruits were spread for curing. Wet and freshly-cut wood which does not burn well and gives a lot of smoke is used as fuel in the kiln so that the temperature does not rise very high because for good curing low temperature is required. This smoking would continue upto three days for complete curing. Approximately 25 per cent of fresh-weight of the fruit is recovered after curing. The colour of capsules remains dark pink-brown and cannot be stored for a longer time.

Now a days flue-pipe system is used which has been found better than the kiln system since the colour of capsules remains shiny-pink. In this system, flue from the furnace passes through the pipes. The harvested wet cardamom is spread over wire meshes fixed above the flue pipes. The product is dried by the movement of hot air. It takes about 24-30 hours for the product to dry. Cardamom produced by this system retains its colour and flavour which fetches good price in the market. Recently an improved version of the flue curing system has been devised by CFTRI, Mysore that is known as natural convection (NC) dryer. The flue ducts can be easily dismantled and transported to other plantations. It has a drying area of 9 m^2 with a capacity of 300 kg raw-capsules per shift which normally takes 24 hours of continuously burning the fuel. Efficiency is also better with a ratio of 2.75:1 of fuel to raw-capsule by weight. But such advance system is not available in the region. Hence it is necessary to adopt an appropriate curing technique which involves indirect heating with optimum temperature range (50-55 °C) when other states of the region like Arunachal, Nagaland, Manipur and Meghalaya are also going to cultivate large cardamom at commercial scale.

Recently passion-fruit production is being promoted on commercial scale in many states like, Manipur, Meghalaya, Mizoram, Nagaland and Sikkim. The concerned state governments are going to install the processing units for production of concentrated juice, squash etc. However, processing plants have already have been installed in Manipur, Mizoram, Sikkim and Nagaland. For making a processing unit viable and profitable, raw material should be made available in such a way that it runs for at least 10 months in a year. This is possible by crop and product diversification, which is possible in the region.

The other units installed in the region suffer heavy losses every year and capacity utilization in public sector is much below the installed capacity. The cost of production is high due to the use of inefficient production processes, machines, excess man-power, poor management systems as well as high over-head costs. The cost of packaging material including cans, bottles, labels, cartons etc is around 35 to 40 per cent of the variable production cost and 20 to 30 per cent of the total production cost because all these materials are procured from outside the region. The cost of transportation to the markets outside the region adds to the cost of the finished products.

Marketing

Marketing is an important component of horticultural development. With the gradual displacement of subsistence farming by commercialized agriculture, marketing of farm produce has assumed greater importance in recent years. Unless the marketing efficiency improves, the commercialization of horticulture will not gather momentum. The marketing system of north-eastern region is not well developed. The major bottleneck is poor road/rail network because at present there is only one national highway and one rail link joining it to other parts of the country. The region needs to be suitably managed to find market avenues leading to proper remuneration to the farmers. Though the region has an added advantage of off-season vegetable production in hilly areas, it can fetch higher prices in the 'main land' when there is scarcity of these vegetables.

For some horticultural crops, there is marketable surplus but for many other commodities domestic market within the region has also not yet been surveyed or developed. About 95 per cent of total ginger produced in the region is marketable surplus. Similarly, one-lakh tonnes of potato have been shown as marketable surplus. These need to be processed and marketed outside the region. Since the marketing infrastructure has not at all developed in the region, it is necessary that meticulous planning and designing should be done for its development.

Establishment of procurement centres at farm level will reduce the burden of transportation of product by the growers to market. Due to quick disposal of fruits and vegetables further to grading, packaging centers and pre-cooling units, spoilage can be reduced to a great extent.

Constraints

Postharvest Handling and Marketing Constraints

1. Produce-handling and marketing is controlled by middlemen who do not bother about marketing standards, grading and scientific packaging.
2. Postharvest quality control is almost non-existent, resulting in considerable damage and wastage.
3. There is wide gap in the technologies available and those applied.
4. There is absence of pre-cooling and cold storage practices.

5. Long distances between the production and major consumption areas are connected through rough, rugged and poorly maintained road network.
6. Existing facilities in the markets are inadequate.
7. Lack of market intelligence and market information is there
8. Storage facilities are inadequate.

Processing Constraints

1. Absence of processing facilities
2. Non-availability of raw materials
3. High cost of basic packaging raw-materials
4. High transportation cost.
5. Lack of trained manpower in postharvest handling and processing..

Thrust areas

Considering the limited and dwindling land and water resources, slow growth in productivity and ever-increasing population, minimizing postharvest losses is one of the most effective and economical ways of increasing per capita food availability. Postharvest technology has to be location specific. However, the present requirement is to develop need based and market-driven postharvest technology for loss prevention and generation of raw materials of horticultural origin. Postharvest losses could be substantially reduced by adopting improved equipments and processes. Following are some of the thrust areas identified in postharvest technology for value-addition and employment generation which need immediate attention.

1. Survey of existing postharvest practices such as harvesting, storage structure, packaging technique, grading and sorting method and locally available products.
2. Refinement and adoption of the presently available postharvest technology for loss prevention and value-addition to horticultural produce.
3. Development of dryers and storage structures to prevent postharvest losses.

4. Development of storage and ware-housing facilities at the collection centres and terminal markets.
5. Introduction and popularisation of simple and efficient postharvest equipment for cleaning, grading, drying, value-addition, packaging and handling of important horticultural crops grown in the region.
6. Testing of packaging materials and development of low-cost packaging structures using locally available materials for long distance transportation.
7. Establishment of horticultural processing unit for each identified cluster of villages to generate employment and encourage the farmers for entrepreneurship.

References

Anonymous.1991. Annual Report. ICAR Research Complex for NEH Region, Barapani, Meghalaya. pp 100-105.

Annonymous, 2002. Basic Statistics of North Eastern Region 2000.North Eastern Councel, Ministry of Home Affair, GOI, Shillong.

Chadha, K.L. and Pareek, O.P.1993. Fruit Production, Research and Development in India. *In*: Singh, R.B.(ed). Research and Development of Fruits in Asia-Pacific Region, Food and Agriculture Organization of United Nations, Bangkok. pp 99-112.

Chadha, K.L.1995. Status of Postharvest Technology of Fruits. *In*: Proc. *National Seminar on Postharvest Technology of Fruits* held at UAS, Bangalore, August 7-9.pp. 1-6.

Gupta, P.N.; Naqui, A.N.; Misra, L.N.; Sen, T. and Nigam,M.C.1984. Gas chromatographic evaluation of the essential oil of different strains of *Amomum subulatum* Roxb. Growing wild in Sikkim. *Sounderdruck aus Perfumcric Kadmetik*.65(9):523-529.

Singh, Kirti and Seshadri, V.S. 1989. Vegetable Production Research and Development in India. FAO Regional Office, Bangkok, Thailand, pp. 169-188.

Singh, R.K.P.; Satapathy, K.K. and Sheo, Govind 1999. Present status of postharvest Practices for fruits and vegetables in NEH Region. *Agricultural Engineering Today*, 1 2: 50-56.

Index

N

Q

R

T

V

W

Y